Aerodynamics of Wind Turbines
Second Edition

Aerodynamics of Wind Turbines

Second Edition

Martin O. L. Hansen

London • Sterling, VA

Second edition published by Earthscan in the UK and USA in 2008
First edition published by James & James (Science Publishers) Ltd in 2000

Copyright © Martin O. L. Hansen, 2008

ICCS3F9101

ISBN: 978-1-84407-438-9

Typeset by FiSH Books, Enfield
Printed and bound in the UK by TJ International, Padstow
Cover design by Nick Shah

For a full list of publications please contact:

Earthscan
8–12 Camden High Street
London, NW1 0JH, UK
Tel: +44 (0)20 7387 8558
Fax: +44 (0)20 7387 8998
Email: earthinfo@earthscan.co.uk
Web: **www.earthscan.co.uk**

22883 Quicksilver Drive, Sterling, VA 20166-2012, USA

Earthscan publishes in association with the International Institute
for Environment and Development

A catalogue record for this book is available from the British Library

Library of Congress Cataloging-in-Publication Data

Hansen, Martin O. L.
  Aerodynamics of wind turbines / Martin O. L. Hansen. — 2nd ed.
     p. cm.
  ISBN-13: 978-1-84407-438-9 (hardback)
  ISBN-10: 1-84407-438-2 (hardback)
  1. Wind turbines. 2. Wind turbines—Aerodynamics. I. Title.
  TJ828.H35 2007
  621.4'5—dc22
                                        2007011666

The paper used for this book is FSC-certified and
totally chlorine-free. FSC (the Forest Stewardship
Council) is an international network to promote
responsible management of the world's forests.

Mixed Sources
Product group from well-managed
forests and other controlled sources
www.fsc.org Cert no. SGS-COC-2482
© 1996 Forest Stewardship Council

Contents

List of Figures and Tables

Figures

Tables

Box

1

General Introduction to Wind Turbines

Before addressing more technical aspects of wind turbine technology, an attempt is made to give a short general introduction to wind energy. This involves a very brief historical part explaining the development of wind power, as well as a part dealing with economy and wind turbine design. It is by no means the intention to give a full historical review of wind turbines, merely to mention some major milestones in their development and to give examples of the historical exploitation of wind power.

Short Historical Review

The force of the wind can be very strong, as can be seen after the passage of a hurricane or a typhoon. Historically, people have harnessed this force peacefully, its most important usage probably being the propulsion of ships using sails before the invention of the steam engine and the internal combustion engine. Wind has also been used in windmills to grind grain or to pump water for irrigation or, as in The Netherlands, to prevent the ocean from flooding low-lying land. At the beginning of the twentieth century electricity came into use and windmills gradually became wind turbines as the rotor was connected to an electric generator.

The first electrical grids consisted of low-voltage DC cables with high losses. Electricity therefore had to be generated close to the site of use. On farms, small wind turbines were ideal for this purpose and in Denmark Poul la Cour, who was among the first to connect a windmill to a generator, gave a course for 'agricultural electricians'. An example of La Cour's great foresight was that he installed in his school one of the first wind tunnels in the world in order to investigate rotor aerodynamics. Gradually, however, diesel engines and steam turbines took over the production of electricity and only during the two world wars, when the supply of fuel was scarce, did wind power flourish again.

However, even after the Second World War, the development of more efficient wind turbines was still pursued in several countries such as Germany, the US, France, the UK and Denmark. In Denmark, this work was undertaken by Johannes Juul, who was an employee in the utility company SEAS and a former student of la Cour. In the mid 1950s Juul introduced what was later called the Danish concept by constructing the famous Gedser turbine, which had an upwind three-bladed, stall regulated rotor, connected to an AC asynchronous generator running with almost constant speed. With the oil crisis in 1973, wind turbines suddenly became interesting again for many countries that wanted to be less dependent on oil imports; many national research programmes were initiated to investigate the possibilities of utilizing wind energy. Large non-commercial prototypes were built to evaluate the economics of wind produced electricity and to measure the loads on big wind turbines. Since the oil crisis, commercial wind turbines have gradually become an important industry with an annual turnover in the 1990s of more than a billion US dollars per year. Since then this figure has increased by approximately 20 per cent a year.

Why Use Wind Power?

As already mentioned, a country or region where energy production is based on imported coal or oil will become more self-sufficient by using alternatives such as wind power. Electricity produced from the wind produces no CO_2 emissions and therefore does not contribute to the greenhouse effect. Wind energy is relatively labour intensive and thus creates many jobs. In remote areas or areas with a weak grid, wind energy can be used for charging batteries or can be combined with a diesel engine to save fuel whenever wind is available. Moreover, wind turbines can be used for the desalination of water in coastal areas with little fresh water, for instance the Middle East. At windy sites the price of electricity, measured in $/kWh, is competitive with the production price from more conventional methods, for example coal fired power plants.

To reduce the price further and to make wind energy more competitive with other production methods, wind turbine manufacturers are concentrating on bringing down the price of the turbines themselves. Other factors, such as interest rates, the cost of land and, not least, the amount of wind available at a certain site, also influence the production price of the electrical energy generated. The production price is computed as the investment plus the discounted maintenance cost divided by the discounted production measured in kWh over a period of typically 20 years. When the characteristics of a given turbine – the power for a given wind speed, as well as the annual wind distribution – are known, the annual energy production can be estimated at a specific site.

Some of the drawbacks of wind energy are also mentioned. Wind turbines create a certain amount of noise when they produce electricity. In modern wind turbines, manufacturers have managed to reduce almost all mechanical noise and are now working hard on reducing aerodynamic noise from the rotating blades. Noise is an important competition factor, especially in densely populated areas. Some people think that wind turbines are unsightly in the landscape, but as bigger and bigger machines gradually replace the older smaller machines, the actual number of wind turbines will be reduced while still increasing capacity. If many turbines are to be erected in a region, it is important to have public acceptance. This can be achieved by allowing those people living close to the turbines to own a part of the project and thus share the income. Furthermore, noise and visual impact will in the future be less important as more wind turbines will be sited offshore.

One problem is that wind energy can only be produced when nature supplies sufficient wind. This is not a problem for most countries, which are connected to big grids and can therefore buy electricity from the grid in the absence of wind. It is, however, an advantage to know in advance what resources will be available in the near future so that conventional power plants can adapt their production. Reliable weather forecasts are desirable since it takes some time for a coal fired power plant to change its production. Combining wind energy with hydropower would be perfect, since it takes almost no time to open or close a valve at the inlet to a water turbine and water can be stored in the reservoirs when the wind is sufficiently strong.

The Wind Resource

A wind turbine transforms the kinetic energy in the wind to mechanical energy in a shaft and finally into electrical energy in a generator. The maximum available energy, P_{max}, is thus obtained if theoretically the wind speed could be reduced to zero: $P = 1/2 \, \dot{m}V_o^2 = 1/2 \, \rho A V_o^3$ where $\dot{m}$ is the mass flow, V_o is the wind speed, ρ the density of the air and A the area where the wind speed has been reduced. The equation for the maximum available power is very important since it tells us that power increases with the cube of the wind speed and only linearly with density and area. The available wind speed at a given site is therefore often first measured over a period of time before a project is initiated.

In practice one cannot reduce the wind speed to zero, so a power coefficient C_p is defined as the ratio between the actual power obtained and the maximum available power as given by the above equation. A theoretical maximum for C_p exists, denoted by the Betz limit, $C_{Pmax} = 16/27 = 0.593$. Modern wind turbines operate close to this limit, with C_p up to 0.5, and are therefore optimized. Statistics have been given on many different turbines sited in Denmark and as rule of thumb they produce approximately 1000kWh/m²/year. However, the production is very site dependent and the rule of thumb can only be used as a crude estimation and only for a site in Denmark.

Sailors discovered very early on that it is more efficient to use the lift force than simple drag as the main source of propulsion. Lift and drag are the components of the force perpendicular and parallel to the direction of the relative wind respectively. It is easy to show theoretically that it is much more efficient to use lift rather than drag when extracting power from the wind. All modern wind turbines therefore consist of a number of rotating blades looking like propeller blades. If the blades are connected to a vertical shaft, the turbine is called a vertical-axis machine, VAWT, and if the shaft is horizontal, the turbine is called a horizontal-axis wind turbine, HAWT. For commercial wind turbines the mainstream mostly consists of HAWTs; the following text therefore focuses on this type of machine. A HAWT as sketched in Figure 1.1 is described in terms of the rotor diameter, the number of blades, the tower height, the rated power and the control strategy.

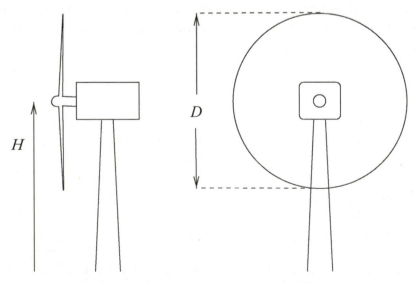

Figure 1.1 *Horizontal-axis wind turbine (HAWT)*

The tower height is important since wind speed increases with height above the ground and the rotor diameter is important since this gives the area A in the formula for the available power. The ratio between the rotor diameter D and the hub height H is often approximately one. The rated power is the maximum power allowed for the installed generator and the control system must ensure that this power is not exceeded in high winds. The number of blades is usually two or three. Two-bladed wind turbines are cheaper since they have one blade fewer, but they rotate faster and appear more flickering to the eyes, whereas three-bladed wind turbines seem calmer and therefore less disturbing in a landscape. The aerodynamic efficiency is lower on a two-bladed than on a three-bladed wind turbine. A two-bladed wind turbine is often, but not always, a downwind machine; in other words the rotor is downwind of the tower. Furthermore, the connection to the shaft is flexible, the rotor being mounted on the shaft through a hinge. This is called a teeter mechanism and the effect is that no bending moments are transferred from the rotor to the mechanical shaft. Such a construction is more flexible than the stiff three-bladed rotor and some components can be built lighter and smaller, which thus reduces the price of the wind turbine. The stability of the more flexible rotor must, however, be ensured. Downwind turbines are noisier than upstream turbines, since the once-per-revolution tower passage of each blade is heard as a low frequency noise.

The rotational speed of a wind turbine rotor is approximately 20 to 50 rpm and the rotational speed of most generator shafts is approximately 1000 to 3000 rpm. Therefore a gearbox must be placed between the low-speed rotor shaft and the high-speed generator shaft. The layout of a typical wind turbine can be seen in Figure 1.2, showing a Siemens wind turbine designed for offshore use. The main shaft has two bearings to facilitate a possible replacement of the gearbox.

This layout is by no means the only option; for example, some turbines are equipped with multipole generators, which rotate so slowly that no gearbox is needed. Ideally a wind turbine rotor should always be perpendicular to the wind. On most wind turbines a wind vane is therefore mounted somewhere on the turbine to measure the direction of the wind. This signal is coupled with a yaw motor, which continuously turns the nacelle into the wind.

The rotor is the wind turbine component that has undergone the greatest development in recent years. The aerofoils used on the first modern wind turbine blades were developed for aircraft and were not optimized for the much higher angles of attack frequently employed by a wind turbine blade. Even though old aerofoils, for instance NACA63-4XX, have been used in the light of experience gained from the first blades, blade manufacturers have

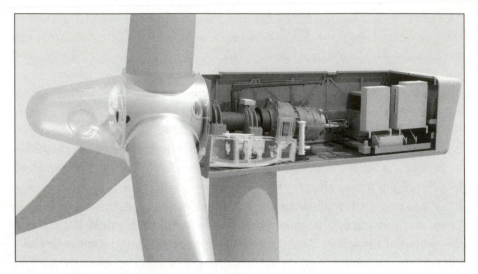

With permission from Siemens Wind Power.

Figure 1.2 *Machine layout*

now started to use aerofoils specifically optimized for wind turbines. Different materials have been tried in the construction of the blades, which must be sufficiently strong and stiff, have a high fatigue endurance limit, and be as cheap as possible. Today most blades are built of glass fibre reinforced plastic, but other materials such as laminated wood are also used.

It is hoped that the historical review, the arguments for supporting wind power and the short description of the technology set out in this chapter will motivate the reader to study the more technical sections concerned with aerodynamics, structures and loads as applied to wind turbine construction.

2

2-D Aerodynamics

Wind turbine blades are long and slender structures where the spanwise velocity component is much lower than the streamwise component, and it is therefore assumed in many aerodynamic models that the flow at a given radial position is two dimensional and that 2-D aerofoil data can thus be applied. Two-dimensional flow is comprised of a plane and if this plane is described with a coordinate system as shown in Figure 2.1, the velocity component in the z-direction is zero.

In order to realize a 2-D flow it is necessary to extrude an aerofoil into a wing of infinite span. On a real wing the chord and twist changes along the span and the wing starts at a hub and ends in a tip, but for long slender wings, like those on modern gliders and wind turbines, Prandtl has shown that local 2-D data for the forces can be used if the angle of attack is corrected accordingly with the trailing vortices behind the wing (see, for example, Prandtl and Tietjens, 1957). These effects will be dealt with later, but it is now clear that 2-D aerodynamics is of practical interest even though it is difficult to realize. Figure 2.1 shows the leading edge stagnation point present in the 2-D flow past an aerofoil. The reacting force $\mathbf{F}$ from the flow is decomposed into a direction perpendicular to the velocity at infinity V_α and to a direction parallel to V_α. The former component is known as the lift, L; the latter is called the drag, D (see Figure 2.2).

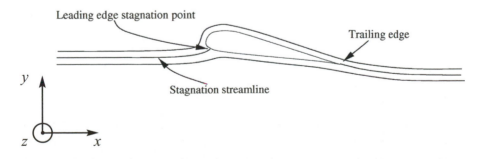

Figure 2.1 *Schematic view of streamlines past an airfoil*

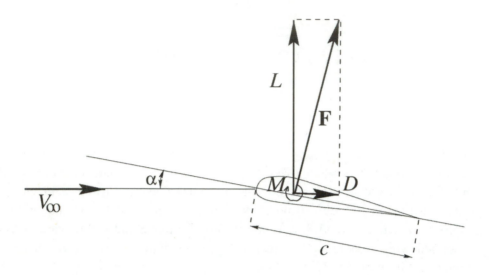

Figure 2.2 *Definition of lift and drag*

If the aerofoil is designed for an aircraft it is obvious that the *L/D* ratio should be maximized. The lift is the force used to overcome gravity and the higher the lift the higher the mass that can be lifted off the ground. In order to maintain a constant speed the drag must be balanced by a propulsion force delivered from an engine, and the smaller the drag the smaller the required engine. Lift and drag coefficients C_l and C_d are defined as:

$$C_l = \frac{L}{1/2\ \rho V\alpha^2 c} \tag{2.1}$$

and:

$$C_d = \frac{D}{1/2\ \rho V\alpha^2 c} \tag{2.2}$$

where ρ is the density and *c* the length of the aerofoil, often denoted by the chord. Note that the unit for the lift and drag in equations (2.1) and (2.2) is force per length (in N/m). A chord line can be defined as the line from the trailing edge to the nose of the aerofoil (see Figure 2.2). To describe the forces completely it is also necessary to know the moment *M* about a point in the aerofoil. This point is often located on the chord line at *c*/4 from the

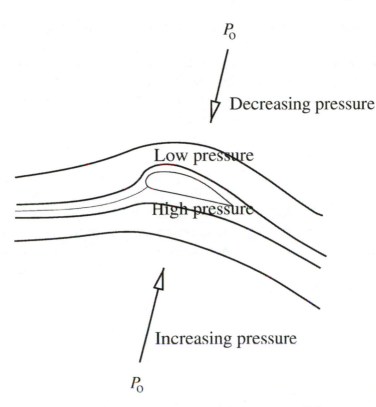

Figure 2.3 *Explanation of the generation of lift*

leading edge. The moment is positive when it tends to turn the aerofoil in Figure 2.2 clockwise (nose up) and a moment coefficient is defined as:

$$C_m = \frac{M}{\frac{1}{2}\rho V_\infty^2 c^2} \tag{2.3}$$

The physical explanation of the lift is that the shape of the aerofoil forces the streamlines to curve around the geometry, as indicated in Figure 2.3. From basic fluid mechanics it is known that a pressure gradient, $\partial p/\partial r = \rho V^2/r$, is necessary to curve the streamlines; r is the curvature of the streamline and V the speed. This pressure gradient acts like the centripetal force known from the circular motion of a particle. Since there is atmospheric pressure p_o far from the aerofoil there must thus be a lower than atmospheric pressure on the upper side of the aerofoil and a higher than atmospheric pressure on the lower side of the aerofoil. This pressure difference gives a lifting force on the aerofoil. When the aerofoil is almost aligned with the flow, the boundary layer stays attached and the associated drag is mainly caused by friction with the air.

The coefficients C_l, C_d and C_m are functions of α, Re and Ma. α is the angle of attack defined as the angle between the chordline and V_∞; Re is the Reynolds number based on the chord and V_∞, Re $= cV_\infty/\nu$, where ν is the kinematic viscosity; and Ma denotes the Mach number, in other words the ratio between V_α and the speed of sound. For a wind turbine and a slow moving aircraft the lift, drag and moment coefficients are only functions of α and Re. For a given airfoil the behaviours of C_l, C_d and C_m are measured or computed and plotted in so-called polars. An example of a measured polar for the FX67-K-170 airfoil is shown in Figure 2.4.

C_l increases linearly with α, with an approximate slope of 2π/rad, until a certain value of α, where a maximum value of C_l is reached. Hereafter the aerofoil is said to stall and C_l decreases in a very geometrically dependent manner. For small angles of attack the drag coefficient C_d is almost constant, but increases rapidly after stall. The Reynolds number dependency can also be seen in Figure 2.4. It is seen, especially on the drag, that as the Reynolds number reaches a certain value, the Reynolds number dependency becomes small. The Reynolds number dependency is related to the point on the aerofoil, where the boundary layer transition from laminar to turbulent flow occurs. The way an aerofoil stalls is very dependent on the geometry. Thin aerofoils with a sharp nose, in other words with high curvature around the leading edge, tend to stall more abruptly than thick aerofoils. Different stall behaviours are seen in Figure 2.5, where $C_l(\alpha)$ is compared for two different aerofoils. The FX38-153 is seen to lose its lift more rapidly than the FX67-K-170.

The explanation lies in the way the boundary layer separates from the upper side of the aerofoil. If the separation starts at the trailing edge of the aerofoil and increases slowly with increasing angle of attack, a soft stall is observed, but if the separation starts at the leading edge of the aerofoil, the entire boundary layer may separate almost simultaneously with a dramatic loss of lift. The behaviour of the viscous boundary layer is very complex and depends, among other things, on the curvature of the aerofoil, the Reynolds number, the surface roughness and, for high speeds, also on the Mach number. Some description of the viscous boundary is given in this text but for a more elaborate description see standard textbooks on viscous boundary layers such as White (1991) and Schlichting (1968).

Figure 2.6 shows the computed streamlines for a NACA63-415 aerofoil at angles of attack of 5° and 15°. For $\alpha = 15°$ a trailing edge separation is observed. The forces on the aerofoil stem from the pressure distribution $p(x)$

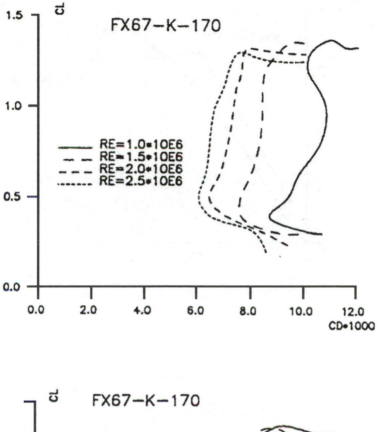

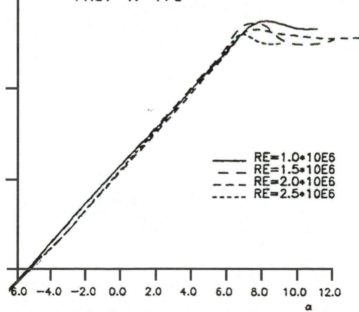

Figure 2.4 *Polar for the FX67-K-170 airfoil*

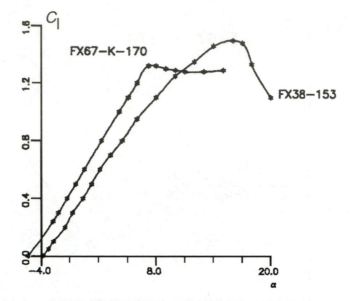

Figure 2.5 *Different stall behaviour*

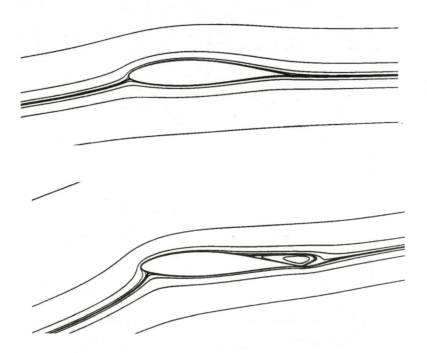

Figure 2.6 *Computed streamlines for angles of attack of 5° and 15°*

and the skin friction with the air $\tau_w = \mu(\partial u/\partial y)_{y=0}$. (x,y) is the surface coordinate system as shown in Figure 2.7 and μ is the dynamic viscosity. The skin friction is mainly contributing to the drag, whereas the force found from integrating the pressure has a lift and drag component. The drag component from the pressure distribution is known as the form drag and becomes very large when the aerofoil stalls. The stall phenomenon is closely related to separation of the boundary layer (see next paragraph); therefore rule number one in reducing drag is to avoid separation. In Abbot and von Doenhoff (1959) a lot of data can be found for the National Advisory Committee for Aeronautics (NACA) aerofoils, which have been extensively used on small aircraft, wind turbines and helicopters. Close to the aerofoil there exists a viscous boundary layer due to the no-slip condition on the velocity at the wall (see Figure 2.7).

A boundary layer thickness is often defined as the normal distance $\delta(x)$ from the wall where $u(x)/U(x) = 0.99$. Further, the displacement thickness $\delta^*(x)$, the momentum thickness $\theta(x)$ and the shape factor $H(x)$ are defined as:

$$\delta^*(x) = \int_0^\delta (1 - \frac{u}{U})dy, \tag{2.4}$$

$$\theta(x) = \int_0^\delta \frac{u}{U}(1 - \frac{u}{U})dy, \text{ and} \tag{2.5}$$

$$H(x) = \frac{\delta^*}{\theta}. \tag{2.6}$$

The coordinate system (x,y) is a local system, where $x = 0$ is at the leading edge stagnation point and y is the normal distance from the wall. A turbulent boundary layer separates for H between 2 and 3. The stagnation streamline (see Figure 2.1) divides the fluid that flows over the aerofoil from the fluid that flows under the aerofoil. At the stagnation point the velocity is zero and the boundary layer thickness is small. The fluid which flows over the aerofoil accelerates as it passes the leading edge and, since the leading edge is close to the stagnation point and the flow accelerates, the boundary layer is thin. It is known from viscous boundary layer theory (see, for example, White, 1991) that the pressure is approximately constant from the surface to the edge of the boundary layer, i.e. $\partial p/\partial y = 0$. Outside the boundary layer the Bernoulli equation (see Appendix A) is valid and, since the flow accelerates, the pressure decreases, i.e. $\partial p/\partial x < 0$. On the lower side the pressure gradient is much smaller since the curvature of the wall is small compared to the leading edge. At the trailing edge the pressure must be the same at the upper and lower side (the Kutta condition) and therefore the pressure must rise, $\partial p/\partial x > 0$, from a minimum value somewhere on the upper side to a higher

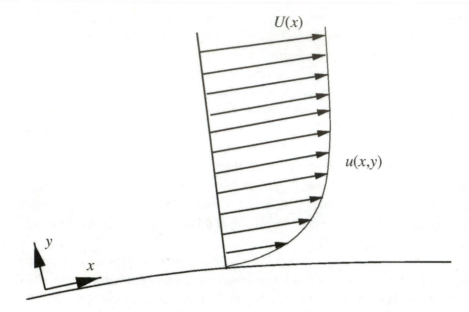

Figure 2.7 *Viscous boundary layer at the wall of an airfoil*

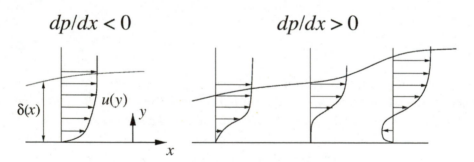

Figure 2.8 *Schematic view of the shape of the boundary layer for a favourable and an adverse pressure gradient*

value at the trailing edge. An adverse pressure gradient, $\partial p/\partial x > 0$, may lead to separation. This can be seen directly from the Navier-Stokes equations (see Appendix A) which applied at the wall, where the velocity is zero, reduces to:

$$\frac{\partial^2 u}{\partial y^2} = \frac{1}{\mu}\frac{\partial p}{\partial x} \tag{2.7}$$

The curvature of the u-velocity component at the wall is therefore given by the sign of the pressure gradient. Further, it is known that $\partial u/\partial y = 0$ at $y = \delta$.

From this can be deduced that the u velocity profile in an adverse pressure gradient, $\partial p/\partial x > 0$, is S-shaped and separation may occur, whereas the curvature of the u velocity profile for $\partial p/\partial x < 0$ is negative throughout the entire boundary layer and the boundary layer stays attached. A schematic picture showing the different shapes of the boundary layer is given in Figure 2.8.

Since the form drag increases dramatically when the boundary layer separates, it is of utmost importance to the performance of an aerofoil to control the pressure gradient.

For small values of x the flow is laminar, but for a certain x_{trans} the laminar boundary layer becomes unstable and a transition from laminar to turbulent flow occurs. At x_{T} the flow is fully turbulent. In Figure 2.9 transition from a laminar to a turbulent boundary layer is sketched. The transitional process is very complex and not yet fully understood, but a description of the phenomena is found in White (1991), where some engineering tools to predict x_{trans} are also given. One of the models which is sometimes used in aerofoil computations is called the one-step method of Michel. The method predicts transition when:

$$\text{Re}_\theta = 2.9 \ \text{Re}_x^{0.4} \tag{2.8}$$

where $\text{Re}_\theta = U(x)\cdot\theta(x)/v$ and $\text{Re}_x = U(x)\cdot x/v$. For a laminar aerofoil (see later), however, the Michel method might be inadequate and more advanced methods such as the e^9 method (see White, 1991) should be applied.

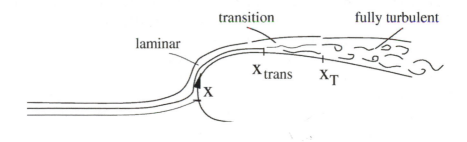

Figure 2.9 *Schematic view of the transitional process*

Turbulent flow is characterized by being more stable in regions of adverse pressure gradients, $\partial p/\partial x > 0$, and by a steeper velocity gradient at the wall, $\partial u/\partial y|_{y=0}$. The first property is good since it delays stall, but the second property increases the skin friction and thus the drag. These two phenomena are exploited in the design of high performance aerofoils called laminar aerofoils. A laminar aerofoil is an aerofoil where a large fraction of the boundary layer is laminar and attached in the range it is designed for. To design such an aerofoil it is necessary to specify the maximum angle of attack, where the boundary layer to a large extent is supposed to be laminar. The aerofoil is then constructed so that the velocity at the edge of the boundary layer, $U(x)$, is constant after the acceleration past the leading edge and downstream. It is known from boundary layer theory (see White, 1991, and Schlichting, 1968) that the pressure gradient is expressed by the velocity outside the boundary layer as:

$$\frac{dp}{dx} = -\rho U(x)\,\frac{dU(x)}{dx} \tag{2.9}$$

At this angle the pressure gradient is therefore zero and no separation will occur. For smaller angles of attack the flow $U(x)$ will accelerate and dp/dx becomes negative, which again avoids separation and is stabilizing for the laminar boundary layer, thus delaying transition. At some point x at the upper side of the aerofoil it is, however, necessary to decelerate the flow in order to fulfil the Kutta condition; in other words the pressure has to be unique at the trailing edge. If this deceleration is started at a position where the boundary layer is laminar, the boundary layer is likely to separate. Just after the laminar/turbulent transition the boundary layer is relatively thin and the momentum close to the wall is relatively large and is therefore capable of withstanding a high positive pressure gradient without separation. During the continuous deceleration towards the trailing edge the ability of the boundary layer to withstand the positive pressure gradient diminishes, and to avoid separation it is therefore necessary to decrease the deceleration towards the trailing edge. It is of utmost importance to ensure that the boundary layer is turbulent before decelerating $U(x)$. To ensure this, a turbulent transition can be triggered by placing a tripwire or tape before the point of deceleration. A laminar aerofoil is thus characterized by a high value of the lift to drag ratio C_l/C_d below the design angle. But before choosing such an aerofoil it is important to consider the stall characteristic and the roughness sensitivity. On an aeroplane it is necessary to fly with a high C_l at landing since the speed is relatively small. If the pilot exceeds $C_{l,max}$ and the aerofoil stalls, it could be disastrous if C_l drops as drastically with the angle of attack as on the FX38-153 in Figure 2.5. The aeroplane would then lose its lift and might slam into the ground.

If the aerofoil is sensitive to roughness, good performance is lost if the wings are contaminated by dust, rain particles or insects, for example. On a wind turbine this could alter the performance with time if, for instance, the turbine is sited in an area with many insects. If a wind turbine is situated near the coast, salt might build up on the blades if the wind comes from the sea, and if the aerofoils used are sensitive to roughness, the power output from the turbine will become dependent on the direction of the wind. Fuglsang and Bak (2003) describe some attempts to design aerofoils specifically for use on wind turbines, where insensitivity to roughness is one of the design targets.

To compute the power output from a wind turbine it is necessary to have data of $C_l(\alpha,Re)$ and $C_d(\alpha,Re)$ for the aerofoils applied along the blades. These data can be measured or computed using advanced numerical tools, but since the flow becomes unsteady and three-dimensional after stall, it is difficult to obtain reliable data for high angles of attack. On a wind turbine very high angles of attack may exist locally, so it is often necessary to extrapolate the available data to high angles of attack.

References

Abbot, H. and von Doenhoff, A. E. (1959) *Theory of Wing Sections,* Dover Publications, New York

Fuglsang, P. and Bak, C. (2003) 'Status of the Risø wind turbine aerofoils', presented at the European Wind Energy Conference, EWEA, Madrid, 16–19 June

Prandtl, L. and Tietjens, O. G. (1957) *Applied Hydro and Aeromechanics,* Dover Publications, New York

Schlichting, H. (1968) *Boundary-Layer Theory,* McGraw-Hill, New York

White, F. M. (1991) *Viscous Fluid Flow,* McGraw-Hill, New York

3

3-D Aerodynamics

This chapter describes qualitatively the flow past a 3-D wing and how the spanwise lift distribution changes the upstream flow and thus the local angle of attack. Basic vortex theory, as described in various textbooks (for example Milne-Thomsen, 1952), is used. Since this theory is not directly used in the Blade Element Momentum method derived later, it is only touched on very briefly here. This chapter may therefore be quite abstract for the reader with limited knowledge of vortex theory, but hopefully some of the basic results will be qualitatively understood.

A wing is a beam of finite length with aerofoils as cross-sections and therefore a pressure difference between the lower and upper sides is created, giving rise to lift. At the tips are leakages, where air flows around the tips from the lower side to the upper side. The streamlines flowing over the wing will thus be deflected inwards and the streamlines flowing under the wing

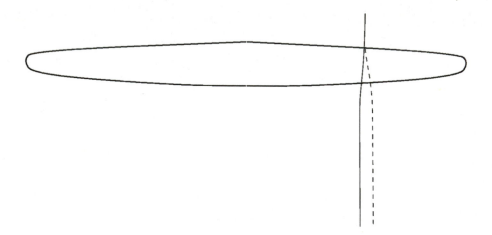

The wing is seen from the suction side. The streamline flowing over the suction side (full line) is deflected inwards and the streamline flowing under (dashed line) is deflected outwards.

Figure 3.1 *Streamlines flowing over and under a wing*

will be deflected outwards. Therefore at the trailing edge there is a jump in the tangential velocity (see Figures 3.1 and 3.2).

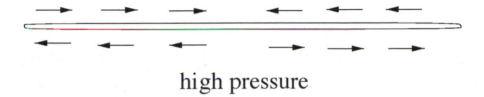

low pressure

high pressure

A jump in the tangential velocity is seen, due to the leakage at the tips.

Figure 3.2 *Velocity vectors seen from behind a wing*

Because of this jump there is a continuous sheet of streamwise vorticity in the wake behind a wing. This sheet is known as the trailing vortices.

In classic literature on theoretical aerodynamics (see, for example, Milne-Thomsen, 1952), it is shown that a vortex filament of strength Γ can model the flow past an aerofoil for small angles of attack. This is because the flow for small angles of attack is mainly inviscid and governed by the linear Laplace equation. It can be shown analytically that for this case the lift is given by the Kutta-Joukowski equation:

$$L = \rho V_\alpha \times \Gamma. \tag{3.1}$$

An aerofoil may be thus substituted by one vortex filament of strength Γ and the lift produced by a 3-D wing can be modelled for small angles of attack by a series of vortex filaments oriented in the spanwise direction of the wing, known as the bound vortices. According to the Helmholtz theorem (Milne-Thomsen, 1952), a vortex filament, however, cannot terminate in the interior of the fluid but must either terminate on the boundary or be closed. A complete wing may be modelled by a series of vortex filaments, Γ_i, i = 1,2,3,4,..., which are oriented as shown in Figure 3.3.

In a real flow the trailing vortices will curl up around the strong tip vortices and the vortex system will look more like that in Figure 3.4.

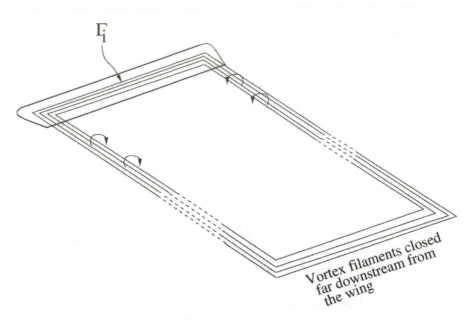

Γ_i

Vortex filaments closed far downstream from the wing

Figure 3.3 *A simplified model of the vortex system on a wing*

The model based on discrete vortices, as shown in Figure 3.3, is called the lifting line theory (see Schlichting and Truckenbrodt, 1959 for a complete description). The vortices on the wing (bound vortices) model the lift, and the trailing vortices (free vortices) model the vortex sheet stemming from the three dimensionality of the wing. The free vortices induce by the Biot-Savart law a downwards velocity component at any spanwise position of the wing. For one vortex filament of strength Γ the induced velocity at a point p is (see Figure 3.5):

$$\mathbf{w} = \frac{\Gamma}{4\pi} \oint \frac{\mathbf{r} \times \mathbf{ds}}{r^3}. \tag{3.2}$$

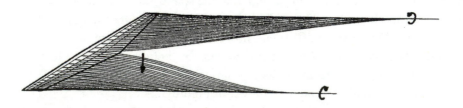

Figure 3.4 *More realistic vortex system on a wing*

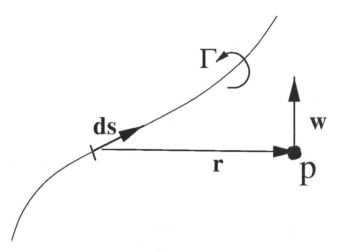

Figure 3.5 *Induced velocity from a vortex line of strength Γ*

The total induced velocity from all vortices at a section of the wing is known as the downwash, and the local angle of attack at this section is therefore reduced by α_i, since the relative velocity is the vector sum of the wind speed V_x and the induced velocity **w**. α_g, α_i and α_e denote the geometric, the induced and the effective angles of attack respectively. The effective angle of attack is thus:

$$\alpha_e = \alpha_g - \alpha_i. \tag{3.3}$$

In Figure 3.6 the induced velocity **w**, the onset flow V_x and the effective velocity V_e are shown for a section on the wing together with the different angles of attack α_g, α_i and α_e. It is assumed that equation (3.1) is also valid for a section in a 3-D wing if the effective velocity is used. The local lift force **R**, which is perpendicular to the relative velocity, is shown in Figure 3.6. The global lift is by definition the force perpendicular to the onset flow V_∞ and the resulting force, **R**, must therefore be decomposed into components perpendicular to and parallel to the direction of V_x. The former component is thus the lift and the latter is a drag denoted by induced drag D_i. At the tips of the wing the induced velocity obtains a value which exactly ensures zero lift.

An important conclusion is thus:
For a three-dimensional wing the lift is reduced compared to a two-dimensional wing at the same geometric angle of attack, and the local lift has a component in the direction of the onset flow, which is known as the induced

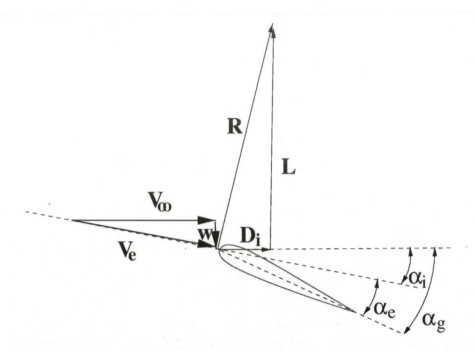

Figure 3.6 *The effective angle of attack for a section in a wing and the resulting force* **R**, *lift* **L** *and induced drag* **D**$_i$

drag. Both effects are due to the downwash induced by the vortex system of a 3-D wing.

In the lifting line theory it is assumed that the three-dimensionality is limited to the downwash, in other words that the spanwise flow is still small compared to the streamwise velocity and 2-D data can therefore be used locally if the geometric angle of attack is modified by the downwash. This assumption is reasonable for long slender wings such as those on a glider or a wind turbine. One method to determine the value of the vortices quantitatively and thus the induced velocities is Multhopp's solution of Prandtl's integral equation. This method is thoroughly described in, for example, Schlichting and Truckenbrodt (1959) and will not be shown here, but it is important to understand that the vortex system produced by a three-dimensional wing changes the local inflow conditions seen by the wing, in other words that although the flow is locally 2-D one cannot apply the geometric angle of attack when estimating the forces on the wing. This error was made in early propeller theory and the discrepancy between measured and computed performance was believed to be caused by wrong 2-D aerofoil data. On a rotating blade Coriolis and centrifugal forces play an important role in the separated boundary layers which occur after stall. In a separated boundary layer the velocity and thus the momentum is relatively small compared to the centrifugal force, which therefore starts to pump fluid in the spanwise direction towards the tip. When the fluid moves radially towards the tip the Coriolis force points towards the trailing edge and acts as a favourable pressure gradient. The effect of the centrifugal and Coriolis force is to alter the 2-D aerofoil data after stall. Considerable engineering skill and experience is required to construct such post-stall data – for example to compute the performance of a wind turbine at high wind speeds – in order to obtain an acceptable result (see also Snel et al, 1993 and Chaviaropoulos and Hansen, 2000).

Figure 3.7 shows the computed limiting streamlines on a modern wind turbine blade at a moderately high wind speed (Hansen et al, 1997). Limiting streamlines are the flow pattern very close to the surface. Figure 3.7 shows that for this specific blade at a wind speed of 10m/s the flow is attached on the outer part of the blade and separated at the inner part, where the limiting streamlines have a spanwise component.

Figure 3.7 *Computed limiting streamlines on a stall regulated wind turbine blade at a moderately high wind speed*

Vortex System behind a Wind Turbine

The rotor of a horizontal-axis wind turbine consists of a number of blades, which are shaped as wings. If a cut is made at a radial distance, *r*, from the rotational axis as shown in Figure 3.8, a cascade of aerofoils is observed as shown in Figure 3.9.

The local angle of attack α is given by the pitch of the aerofoil θ; the axial velocity and rotational velocity at the rotor plane denoted respectively by V_a and V_{rot} (see Figure 3.9):

$$\alpha = \phi - \theta \tag{3.4}$$

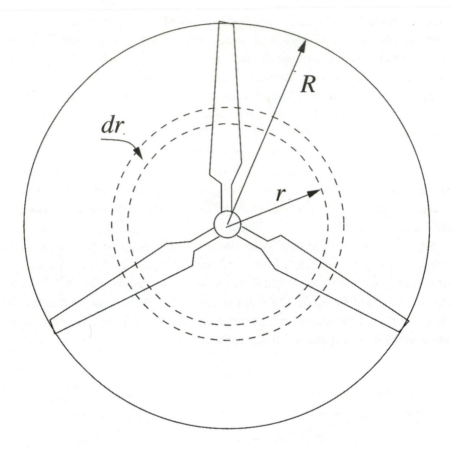

Figure 3.8 *Rotor of a three-bladed wind turbine with rotor radius* R

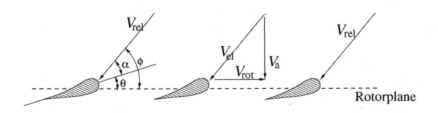

Figure 3.9 *Radial cut in a wind turbine rotor showing airfoils at* r/R

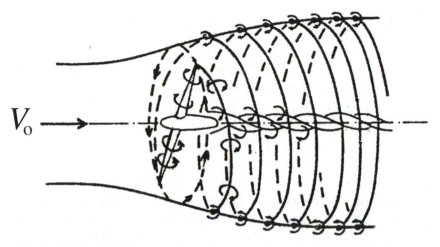

Source: Wilson and Lissaman (1974), reproduced with permission

Figure 3.10 *Schematic drawing of the vortex system behind a wind turbine*

where the flow angle ϕ is found as:

$$\tan \phi = \frac{V_a}{V_{rot}} \tag{3.5}$$

Since a horizontal-axis wind turbine consists of rotating blades, a vortex system similar to the linear translating wing must exist. The vortex sheet of the free vortices is oriented in a helical path behind the rotor. The strong tip vortices are located at the edge of the rotor wake and the root vortices lie mainly in a linear path along the axis of the rotor, as shown in Figure 3.10.

The vortex system induces on a wind turbine an axial velocity component opposite to the direction of the wind and a tangential velocity component opposite to the rotation of the rotor blades. The induced velocity in the axial direction is specified through the axial induction factor a as aV_o, where V_o is the undisturbed wind speed. The induced tangential velocity in the rotor wake is specified through the tangential induction factor a' as $2a'\omega r$. Since the flow does not rotate upstream of the rotor, the tangential induced velocity in the rotor plane is thus approximately $a'\omega r$. ω denotes the angular velocity of the rotor and r is the radial distance from the rotational axis. If a and a' are known, a 2-D equivalent angle of attack could be found from equations (3.4) and (3.5), where:

$$V_a = (1-a)V_o, \tag{3.6}$$

and:

$$V_{rot} = (1 + a')\omega r. \tag{3.7}$$

Furthermore, if the lift and drag coefficients $C_l(\alpha)$ and $C_d(\alpha)$ are also known for the aerofoils applied along the blades, it is easy to compute the force distribution. Global loads such as the power output and the root bending moments of the blades are found by integrating this distribution along the span. It is the purpose of the Blade Element Momentum method, which will later be derived in detail, to compute the induction factors a and a' and thus the loads on a wind turbine. It is also possible to use a vortex method and construct the vortex system as shown in Figure 3.10 and use the Biot-Savart equation (3.2) to calculate the induced velocities. Such methods are not derived in this book but can found in, for example, Katz and Plotkin (2001) or Leishman (2006).

References

Chaviaropoulos, P. K. and Hansen, M. O. L. (2000) 'Investigating three-dimensional and rotational effects on wind turbine blades by means of a quasi-3D Navier-Stokes solver', *Journal of Fluids Engineering*, vol 122, pp330–336

Hansen, M. O. L, Sorensen, J. N., Michelsen, J. A. and Sorensen, N. N. (1997) 'A global Navier-Stokes rotor prediction model', AIAA 97-0970 paper, 35th Aerospace Sciences Meeting and Exhibition, Reno, Nevada, 6–9 January

Katz, J. and Plotkin, A. (2001) *Low-Speed Aerodynamics*, Cambridge University Press, Cambridge

Leishmann, J. G. (2006) *Principles of Helicopter Aerodynamics*, Cambridge University Press, Cambridge

Milne-Thomson, L. M. (1952) *Theoretical Aerodynamics*, Macmillan, London

Schlichting, H. and Truckenbrodt, E. (1959) *Aerodynamik des Flugzeuges*, Springer-Verlag, Berlin

Snel, H., Houwink, B., Bosschers, J., Piers, W. J., van Bussel, G. J. W. and Bruining, A. (1993) 'Sectional prediction of 3-D effects for stalled flow on rotating blades and comparison with measurements', in *Proceedings of European Community Wind Energy Conference 1993*, H. S. Stephens & Associates, Travemunde, pp395–399

Wilson, R. E. and Lissaman, P. B. S. (1974) *Applied Aerodynamics of Wind Power Machines*, Technical Report NSF-RA-N-74-113, Oregon State University

4

1-D Momentum Theory for an Ideal Wind Turbine

Before deriving the Blade Element Momentum method it is useful to examine a simple one-dimensional (1-D) model for an ideal rotor. A wind turbine extracts mechanical energy from the kinetic energy of the wind. The rotor in this simple 1-D model is a permeable disc. The disc is considered ideal; in other words it is frictionless and there is no rotational velocity component in the wake. The latter can be obtained by applying two contra-rotating rotors or a stator. The rotor disc acts as a drag device slowing the wind speed from V_o far upstream of the rotor to u at the rotor plane and to u_l in the wake. Therefore the streamlines must diverge as shown in Figure 4.1. The drag is obtained by a pressure drop over the rotor. Close upstream of the rotor there is a small pressure rise from the atmospheric level p_o to p before a discontinuous pressure drop Δp over the rotor. Downstream of the rotor the pressure recovers continuously to the atmospheric level. The Mach number is small and the air density is thus constant and the axial velocity must decrease continuously from V_o to u_l. The behaviour of the pressure and axial velocity is shown graphically in Figure 4.1.

Using the assumptions of an ideal rotor it is possible to derive simple relationships between the velocities V_o, u_l and u, the thrust T, and the absorbed shaft power P. The thrust is the force in the streamwise direction resulting from the pressure drop over the rotor, and is used to reduce the wind speed from V_o to u_l:

$$T = \Delta p A, \tag{4.1}$$

where $A = \pi R^2$ is the area of the rotor. The flow is stationary, incompressible and frictionless and no external force acts on the fluid up- or downstream of the rotor. Therefore the Bernoulli equation (see Appendix A) is valid from far upstream to just in front of the rotor and from just behind the rotor to far downstream in the wake:

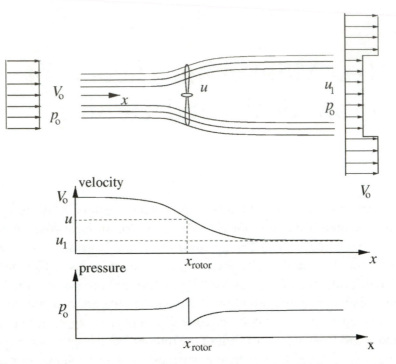

Figure 4.1 *Illustration of the streamlines past the rotor and the axial velocity and pressure up- and downstream of the rotor*

$$p_\text{o} + \frac{1}{2}\rho V_\text{o}^2 = p + \frac{1}{2}\rho u^2, \tag{4.2}$$

and:

$$p - \Delta p + \frac{1}{2}\rho u^2 = p_\text{o} + \frac{1}{2}\rho u_1^2. \tag{4.3}$$

Combining equation (4.2) and (4.3) yields:

$$\Delta p = \frac{1}{2}\rho(V_\text{o}^2 - u_1^2). \tag{4.4}$$

The axial momentum equation in integral form (see Appendix A) is applied on the circular control volume with sectional area A_{cv} drawn with a dashed line in Figure 4.2 yielding:

$$\frac{\partial}{\partial t}\iiint_{cv}\rho u(x, y, z)dxdydz + \iint_{cs} u(x, y, z)\rho \mathbf{V}\cdot\mathbf{dA} = F_{ext} + F_{pres}. \tag{4.5}$$

dA is a vector pointing outwards in the normal direction of an infinitesimal part of the control surface with a length equal to the area of this element. F_{pres} is the axial component of the pressure forces acting on the control volume. The first term in equation (4.5) is zero since the flow is assumed to be stationary and the last term is zero since the pressure has the same atmospheric value on the end planes and acts on an equal area. Further, on the lateral boundary of the control volume shown in Figure 4.2, the force from the pressure has no axial component.

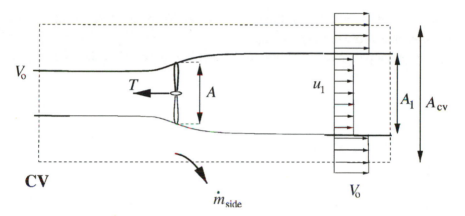

Figure 4.2 *Circular control volume around a wind turbine*

Using the simplified assumptions of an ideal rotor, equation (4.5) then yields:

$$\rho u_1^2 A_1 + \rho V_o^2 (A_{cv} - A_1) + \dot{m}_{side} V_o - \rho V_o^2 A_{cv} = -T. \tag{4.6}$$

$\dot{m}_{side}$ can be found from the conservation of mass:

$$\rho A_1 u_1 + \rho(A_{cv} - A_1)V_o + \dot{m}_{side} = \rho A_{cv} V_o. \tag{4.7}$$

yielding:

$$\dot{m}_{side} = \rho A_1(V_o - u_1). \tag{4.8}$$

The conservation of mass also gives a relationship between A and A_1 as:

$$\dot{m} = \rho u A = \rho u_1 A_1. \tag{4.9}$$

Combining equations (4.8), (4.9) and (4.6) yields:

$$T = \rho u A(V_o - u_1) = \dot{m}(V_o - u_1).\tag{4.10}$$

If the thrust is replaced by the pressure drop over the rotor as in equation (4.1) and the pressure drop from equation (4.4) is used, an interesting observation is made:

$$u = \frac{1}{2}(V_o + u_1).\tag{4.11}$$

It is seen that the velocity in the rotor plane is the mean of the wind speed V_o and the final value in the wake u_1.

An alternative control volume to the one in Figure 4.2 is shown in Figure 4.3.

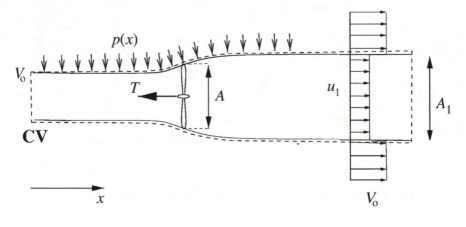

Figure 4.3 *Alternative control volume around a wind turbine*

The force from the pressure distribution along the lateral walls $F_{press,\,lateral}$ of the control volume is unknown and thus so is the net pressure contribution F_{pres}. On this alternative control volume there is no mass flow through the lateral boundary, since this is aligned with the streamlines. The axial momentum equation (4.5) therefore becomes:

$$T = \rho u A(V_o - u_1) + F_{pres}.\tag{4.12}$$

Since the physical problem is the same, whether the control volume in Figure 4.2 or that in Figure 4.3 is applied, it can be seen by comparing equations

(4.10) and (4.12) that the net pressure force on the control volume following the streamlines is zero.

The flow is assumed to be frictionless and there is therefore no change in the internal energy from the inlet to the outlet and the shaft power P can be found using the integral energy equation on the control volume shown in Figure 4.3:

$$P = \dot{m} \left(\frac{1}{2} V_o^2 + \frac{P_o}{\rho} - \frac{1}{2} u_1^2 - \frac{P_o}{\rho} \right). \tag{4.13}$$

and since $\dot{m} = \rho u A$ the equation for P simply becomes:

$$P = \frac{1}{2} \rho u A (V_o^2 - u_1^2). \tag{4.14}$$

The axial induction factor a is defined as:

$$u = (1 - a)V_o. \tag{4.15}$$

Combining equation (4.15) with (4.11) gives:

$$u_1 = (1 - 2a)V_o, \tag{4.16}$$

which then can be introduced in equation (4.14) for the power P and into equation (4.10) for the thrust T, yielding:

$$P = 2\rho V_o^3 a (1 - a)^2 A \tag{4.17}$$

and:

$$T = 2\rho V_o^2 a (1 - a) A. \tag{4.18}$$

The available power in a cross-section equal to the swept area A by the rotor is:

$$P_{avail} = \frac{1}{2} \rho A V_o^3 \tag{4.19}$$

The power P is often non-dimensionalized with respect to P_{avail} as a power coefficient C_p:

$$C_p = \frac{P}{\frac{1}{2} \rho V_o^3 A} \tag{4.20}$$

Similarly a thrust coefficient C_T is defined as:

$$C_T = \frac{T}{\frac{1}{2}\rho V_o^2 A} \tag{4.21}$$

Using equations (4.17) and (4.18) the power and thrust coefficients for the ideal 1-D wind turbine may be written as:

$$C_p = 4a(1-a)^2 \tag{4.22}$$

and:

$$C_T = 4a(1-a). \tag{4.23}$$

Differentiating C_p with respect to a yields:

$$\frac{dC_p}{da} = 4(1-a)(1-3a). \tag{4.24}$$

It is easily seen that $C_{p,\,max} = 16/27$ for $a = 1/3$. Equations (4.22) and (4.23) are shown graphically in Figure 4.4. This theoretical maximum for an ideal wind turbine is known as the Betz limit.

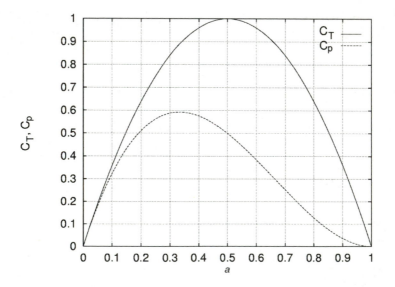

Figure 4.4 *The power and thrust coefficients C_p and C_T as a function of the axial induction factor a for an ideal horizontal-axis wind turbine*

Experiments have shown that the assumptions of an ideal wind turbine leading to equation (4.23) are only valid for an axial induction factor, *a*, of less than approximately 0.4. This is seen in Figure 4.5, which shows measurements of C_T as a function of *a* for different rotor states. If the momentum theory were valid for higher values of *a*, the velocity in the wake would become negative as can readily be seen by equation (4.16).

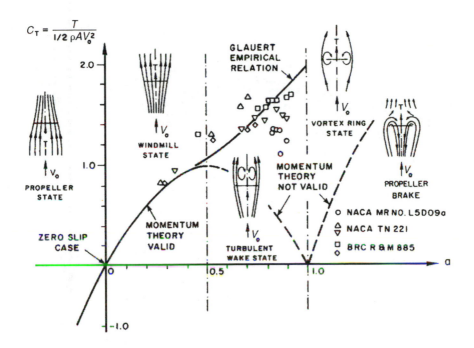

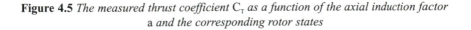

Source: Eggleston and Stoddard (1987), reproduced with permission.

Figure 4.5 *The measured thrust coefficient* C_T *as a function of the axial induction factor* a *and the corresponding rotor states*

As C_T increases the expansion of the wake increases and thus also the velocity jump from V_o to u_1 in the wake, see Figure 4.6.

The ratio between the areas A_o and A_1 in Figure 4.6 can be found directly from the continuity equation as:

$$\frac{A_o}{A_1} = 1 - 2a. \tag{4.25}$$

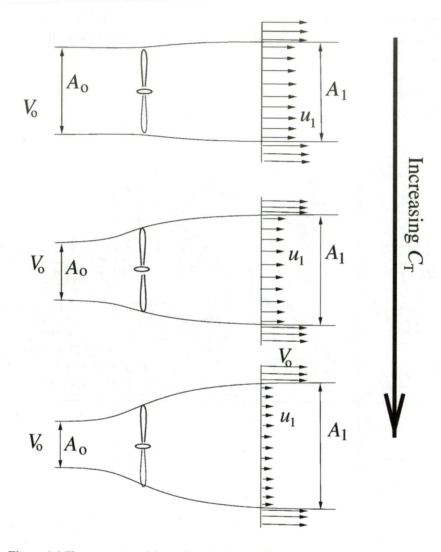

Figure 4.6 *The expansion of the wake and the velocity jump in the wake for the 1-D model of an ideal wind turbine*

For a wind turbine, a high thrust coefficient C_T, and thus a high axial induction factor a, is present at low wind speeds. The reason that the simple momentum theory is not valid for values of a greater than approximately 0.4 is that the free shear layer at the edge of the wake becomes unstable when the velocity jump $V_o - u_l$ becomes too high and eddies are formed which transport momentum from the outer flow into the wake. This situation is called the turbulent-wake state, see Figures 4.5 and 4.7.

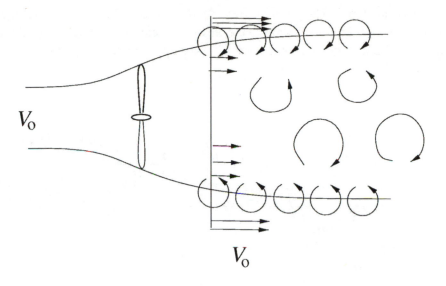

Figure 4.7 *Schematic view of the turbulent-wake state induced by the unstable shear flow at the edge of the wake*

Effects of Rotation

For the ideal rotor there is no rotation in the wake; in other words a' is zero. Since a modern wind turbine consists of a single rotor without a stator, the wake will possess some rotation as can be seen directly from Euler's turbine equation (see Appendix A) applied to an infinitesimal control volume of thickness dr, as shown in Figure 3.8:

$$dP = \dot{m}\omega r \, C_\theta = 2\pi r^2 \, \rho u \omega C_\theta dr, \qquad (4.26)$$

where C_θ is the azimuthal component of the absolute velocity $\mathbf{C} = (C_r, C_\theta, C_a)$ after the rotor and u the axial velocity through the rotor.

Since the forces felt by the wind turbine blades are also felt by the incoming air, but with opposite sign, the air at a wind turbine will rotate in the opposite direction from that of the blades. This can also be illustrated using Figure 4.8, where the relative velocity upstream of the blade $V_{rel,1}$ is given by the axial velocity u and the rotational velocity V_{rot}. For moderate angles of attack the relative velocity $V_{rel,2}$ downstream of the rotor approximately follows the trailing edge. The axial component, C_a, of the absolute velocity equals u due to conservation of mass, and the rotational speed is unaltered. The velocity triangle downstream of the blade is now fixed and, as Figure 4.8 shows, the absolute velocity downstream of the blade, **C**, has a tangential component C_θ in the opposite direction of the blade.

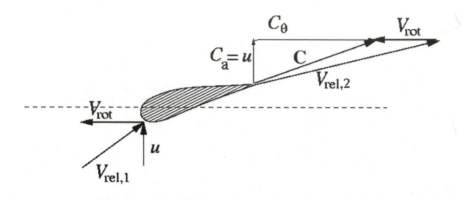

Figure 4.8 *The velocity triangle for a section of the rotor*

From equation (4.26) it is seen that for a given power P and wind speed the azimuthal velocity component in the wake C_θ decreases with increasing rotational speed ω of the rotor. From an efficiency point of view it is therefore desirable for the wind turbine to have a high rotational speed to minimize the loss of kinetic energy contained in the rotating wake. If we recall that the axial velocity through the rotor is given by the axial induction factor a as in equation (4.15) and that the rotational speed in the wake is given by a' as:

$$C_\theta = 2a'\omega r. \tag{4.27}$$

Equation (4.26) may then be written as:

$$dP = 4\pi\rho\omega^2 V_o a'(1 - a)r^3 dr. \tag{4.28}$$

The total power is found by integrating dP from 0 to R as:

$$P = 4\pi\rho\omega^2 V_o \int_0^R a'(1 - a)r^3 dr. \tag{4.29}$$

or in non-dimensional form as:

$$C_p = \frac{8}{\lambda^2} \int_0^\lambda a'(1-a)x^3 dx, \tag{4.30}$$

where $\lambda = \omega R/V_o$ is the tip speed ratio and $x = \omega r/V_o$ is the local rotational speed at the radius r non-dimensionalized with respect to the wind speed V_o. It is clear from equations (4.29) and (4.30) that in order to optimize the power it is necessary to maximize the expression:

$$f(a, a') = a'(1-a). \tag{4.31}$$

If the local angles of attack are below stall, a and a' are not independent since the reacting force according to potential flow theory is perpendicular to the local velocity seen by the blade as indicated by equation (3.1). The total induced velocity, $\mathbf{w}$, must be in the same direction as the force and thus also perpendicular to the local velocity. The following relationship therefore exists between a and a':

$$x^2 a'(1 + a') = a(1 - a). \tag{4.32}$$

Equation (4.32) is directly derived from Figure 4.9 since:

$$\tan \phi = \frac{a'\omega r}{aV_o} \tag{4.33}$$

and:

$$\tan \phi = \frac{(1-a)V_o}{(1+a')\omega r} \tag{4.34}$$

$x = \omega r/V_o$ denotes the ratio between the local rotational speed and the wind speed.

For local angles of attack below stall a and a' are related through equation (4.32) and the optimization problem is thus to maximize equation (4.31) and still satisfy equation (4.32). Since a' is a function of a, the expression (4.31) is maximum when $df/da = 0$ yielding:

$$\frac{df}{da} = (1-a)\frac{da'}{da} - a' = 0, \tag{4.35}$$

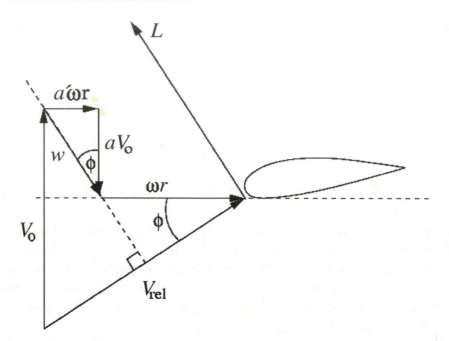

Note that for small angles of attack the total induced velocity w is perpendicular to the relative velocity.

Figure 4.9 *Velocity triangle showing the induced velocities for a section of the blade*

which can be simplified to:

$$(1-a)\frac{da'}{da}=a' \tag{4.36}$$

Equation (4.32) differentiated with respect to a yields:

$$(1+2a')\frac{da'}{da}x^2=1-2a. \tag{4.37}$$

If equations (4.36) and (4.37) are combined with equation (4.32), the optimum relationship between a and a' becomes:

$$a'=\frac{1-3a}{4a-1}. \tag{4.38}$$

A table between a, a' and x can now be computed. a' is given by equation (4.38) for a specified a and then x is found using (4.32).

It can be seen that as the rotational speed ω and thus a $\circ$ $x = \omega r/V_o$ is increased the optimum value for a tends to 1/3, which is co $\;$ stent with the simple momentum theory for an ideal rotor. Using the values from the table, the optimum power coefficient C_p is found by integrating equation (4.30). This is done in Glauert (1935) for different tip speed ratios $\lambda = \omega R/V_o$. Glauert compares this computed optimum power coefficient with the Betz limit of 16/27, which is derived for zero rotation in the wake $a' = 0$ (see Table 4.2). In Figure 4.10, Table 4.2 is plotted and it can be seen that the loss due to rotation is small for tip speed ratios greater than approximately 6.

Table 4.1 *The numerical relationships between a, a' and x*

a	a'	x
0.26	5.5	0.073
0.27	2.375	0.157
0.28	1.333	0.255
0.29	0.812	0.374
0.30	0.500	0.529
0.31	0.292	0.753
0.32	0.143	1.15
0.33	0.031	2.63
0.333	0.00301	8.58

Table 4.2 *Glauert's comparison of the computed optimum power coefficient including wake rotation with the Betz limit*

$\lambda = \omega R/V_o$	$27C_p/16$
0.5	0.486
1.0	0.703
1.5	0.811
2.0	0.865
2.5	0.899
5.0	0.963
7.5	0.983
10.0	0.987

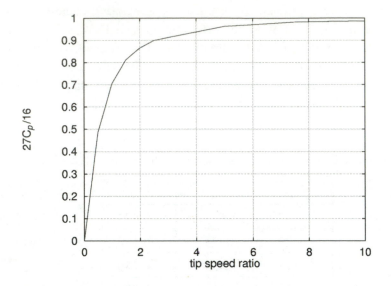

The efficiency is defined as the ratio between C_p, including wake rotation, and the Betz limit $C_{p,\,Betz} = 16/27$.

Figure 4.10 *The efficiency of an optimum turbine with rotation*

References

Eggleston, D. M. and Stoddard, F. S. (1987) *Wind Turbine Engineering Design*, Van Nostrand Reinhold Company, New York

Glauert, H. (1935) 'Airplane propellers', in W. F. Durand (ed) *Aerodynamic Theory*, vol 4, Division L, Julius Springer, Berlin, pp169–360

5

Shrouded Rotors

It is possible to exceed the Betz limit by placing the wind turbine in a diffuser. If the cross-section of the diffuser is shaped like an aerofoil, a lift force will be generated by the flow through the diffuser as seen in Figure 5.1.

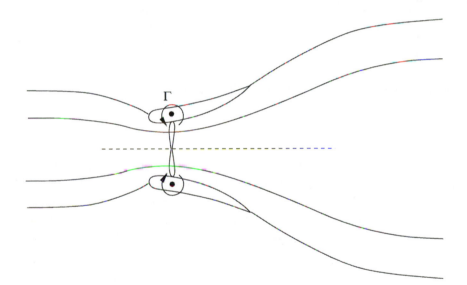

Figure 5.1 *Ideal flow through a wind turbine in a diffuser*

As shown in de Vries (1979), the effect of this lift is to create a ring vortex, which by the Biot-Savart law will induce a velocity to increase the mass flow through the rotor. The axial velocity in the rotor plane is denoted by V_2, and ε is the augmentation defined as the ratio between V_2 and the wind speed V_o, i.e. $\varepsilon = V_2/V_o$. A 1-D analysis of a rotor in a diffuser gives the following expression for the power coefficient:

$$C_{p,d} = \frac{P}{\frac{1}{2}\rho V_o^3 A} = \frac{T \cdot V_2}{\frac{1}{2}\rho V_o^2 \frac{V_o}{V_2} V_2 A} = C_T \varepsilon. \tag{5.1}$$

For an ideal bare turbine equations (4.22) and (4.23) are valid yielding:

$$C_{p,b} = C_T(1 - a). \tag{5.2}$$

Combining equations (5.1) and (5.2) yields:

$$\frac{C_{p,d}}{C_{p,b}} = \frac{\varepsilon}{(1 - a)} \tag{5.3}$$

Further, the following equations are valid for the mass flow through a bare turbine $\dot{m}_b$ and the mass flow through a turbine in a diffuser $\dot{m}_d$:

$$\frac{\dot{m}_b}{\rho V_o A} = \frac{\rho(1 - a)V_o A}{\rho V_o A} = 1 - a \tag{5.4}$$

$$\frac{\dot{m}_d}{\rho V_o A} = \frac{\rho V_2 A}{\rho V_o A} = \varepsilon. \tag{5.5}$$

Combining equations (5.3), (5.4) and (5.5) yields:

$$\frac{C_{p,d}}{C_{p,b}} = \frac{\dot{m}_d}{\dot{m}_b} \tag{5.6}$$

Equation (5.6) states that the relative increase in the power coefficient for a shrouded turbine is proportional to the ratio between the mass flow through the turbine in the diffuser and the same turbine without the diffuser. Equation (5.6) is verified by computational fluid dynamics (CFD) results as seen in Figure 5.2, where for a given geometry the computed mass flow ratio $\dot{m}_d/\dot{m}_b$ is plotted against the computed ratio $C_{p,d}/C_{p,b}$. The CFD analysis is done on a simple geometry, without boundary layer bleed slots, as suggested by Gilbert and Foreman (1983). The diffuser was modelled using 266,240 grid points with 96 points around the diffuser aerofoil section, and a turbulence model was chosen which is sensitive to adverse pressure gradients (see Hansen et al, 2000). The rotor was modelled by specifying a constant volume force at the position of the rotor.

To check this approach some initial computations were made without the diffuser and in Figure 5.3, which shows the relationship between the thrust and power coefficients, it is seen that this approach gave good results compared to the following theoretical expression, which can be derived from equations (4.22) and (4.23):

$$C_{p,b} = \frac{1}{2}C_T (1 + \sqrt{1 - C_T}) \tag{5.7}$$

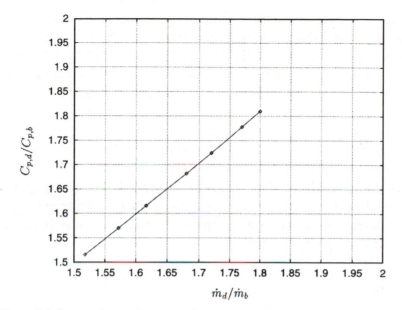

Figure 5.2 *Computed mass flow ratio plotted against the computed power coefficient ratio for a bare and a shrouded wind turbine, respectively*

In Figure 5.3 it is also seen that computations with a wind turbine in a diffuser gave higher values for the power coefficient than the Betz limit for a bare turbine.

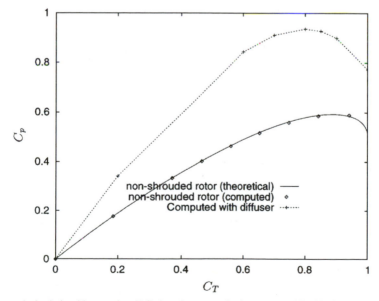

The theoretical relationship equation (5.7) for a bare rotor is also compared in this figure.

Figure 5.3 *Computed power coefficient for a rotor in a diffuser as a function of the thrust coefficient C_T*

The results are dependent on the actual diffuser geometry, in other words on the amount of lift which can be generated by the diffuser. An adverse pressure gradient is present for the flow in the diffuser, and the boundary layer will separate if the ratio between the exit area and the area in the diffuser becomes too high. To increase the lift, giving a higher mass flow through the turbine and thus a higher power output, any trick to help prevent the boundary layer from separating is allowed, for example vortex generators or boundary bleed slots. The computations in Figure 5.3 and the wind tunnel measurements of Gilbert and Foreman (1983) show that the Betz limit can be exceeded if a device increasing the mass flow through the rotor is applied, but this still has to be demonstrated on a full size machine. Furthermore, the increased energy output has to be compared to the extra cost of building a diffuser and the supporting structure.

References

Gilbert, B. L. and Foreman, K. M. (1983) 'Experiments with a diffuser-augmented model wind turbine', *Journal of Energy Resources Technology*, vol 105, pp46–53

Hansen, M. O. L, Sorensen, N. N. and Flay, R. G. J. (2000) 'Effect of placing a diffuser around a wind turbine', *Wind Energy*, vol 3, pp207–213

de Vries, O. (1979) *Fluid Dynamic Aspects of Wind Energy Conversion*, AGARDograph No 243, Advisory Group for Aeronautical Research and Development

6

The Classical Blade Element Momentum Method

All definitions and necessary theory to understand the Blade Element Momentum (BEM) method have now been introduced. In this chapter the classical BEM model from Glauert (1935) will be presented. With this model it is possible to calculate the steady loads and thus also the thrust and power for different settings of wind speed, rotational speed and pitch angle. To calculate time series of the loads for time-varying input some engineering models must be added, as will be shown in a later chapter. In the 1-D momentum theory the actual geometry of the rotor – the number of blades, the twist and chord distribution, and the aerofoils used – is not considered. The Blade Element Momentum method couples the momentum theory with the local events taking place at the actual blades. The stream tube introduced in the 1-D momentum theory is discretized into N annular elements of height dr, as shown in Figure 6.1. The lateral boundary of these elements consists of streamlines; in other words there is no flow across the elements.

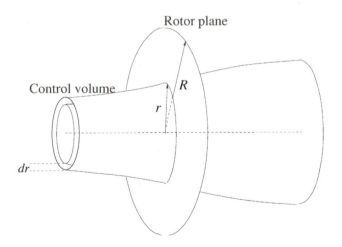

Figure 6.1 *Control volume shaped as an annular element to be used in the BEM model*

In the BEM model the following is assumed for the annular elements:

1 No radial dependency – in other words what happens at one element cannot be felt by the others.

2 The force from the blades on the flow is constant in each annular element; this corresponds to a rotor with an infinite number of blades.

A correction known as Prandtl's tip loss factor is later introduced to correct for the latter assumption in order to compute a rotor with a finite number of blades.

In the previous section concerning the 1-D momentum theory it was proven that the pressure distribution along the curved streamlines enclosing the wake does not give an axial force component. Therefore it is assumed that this is also the case for the annular control volume shown in Figure 6.1. The thrust from the disc on this control volume can thus be found from the integral momentum equation since the cross-section area of the control volume at the rotor plane is $2\pi r dr$:

$$dT = (V_o - u_1)d\dot{m} = 2\pi r \rho u(V_o - u_1)dr. \tag{6.1}$$

The torque dM on the annular element is found using the integral moment of momentum equation on the control volume (see Appendix A) and setting the rotational velocity to zero upstream of the rotor and to C_θ in the wake:

$$dM = rC_\theta d\dot{m} = 2\pi r^2 \rho u C_\theta dr. \tag{6.2}$$

This could also have been derived directly from Euler's turbine equation (4.26), since:

$$dP = \omega dM \tag{6.3}$$

From the ideal rotor it was found that the axial velocity in the wake u_1 could be expressed by the axial induction factor a and the wind speed V_o as $u_1 = (1 - 2a)V_o$, and if this is introduced into equations (6.1) and (6.2) together with the definitions for a and a' in equations (4.15) and (4.27) the thrust and torque can be computed as:

$$dT = 4\pi r \rho V_o^2 \, a(1 - a)dr \tag{6.4}$$

and:

$$dM = 4\pi r^3 \rho V_o \omega (1 - a) a' dr. \tag{6.5}$$

The left hand sides of equations (6.4) and (6.5) are found from the local flow around the blade. It is recalled that the relative velocity V_{rel} seen by a section of the blade is a combination of the axial velocity $(1 - a)V_o$ and the tangential velocity $(1 + a')\omega r$ at the rotorplane (see Figure 6.2).

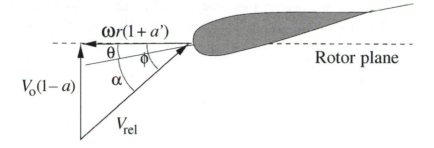

Figure 6.2 *Velocities at the rotor plane*

θ is the local pitch of the blade, in other words the local angle between the chord and the plane of rotation. The local pitch is a combination of the pitch angle, θ_p, and the twist of the blade, β, as $\theta = \theta_p + \beta$, where the pitch angle is the angle between the tip chord and the rotorplane and the twist is measured relative to the tip chord. ϕ is the angle between the plane of rotation and the relative velocity, V_{rel}, and it is seen in Figure 6.2 that the local angle of attack is given by:

$$\alpha = \phi - \theta. \tag{6.6}$$

Further, it is seen that:

$$\tan \phi = \frac{(1 - a)V_o}{(1 + a')\omega r}. \tag{6.7}$$

It is recalled from the section concerning 2-D aerodynamics that the lift, by definition, is perpendicular to the velocity seen by the aerofoil and the drag is parallel to the same velocity. In the case of a rotor this velocity is V_{rel} due to arguments given in the section about the vortex system of a wind turbine.

Further, if the lift and drag coefficients C_l and C_d are known, the lift L and drag D force per length can be found from equations (2.1) and (2.2):

$$L = \frac{1}{2}\rho V_{rel}^2 cC_l \tag{6.8}$$

and:

$$D = \frac{1}{2}\rho V_{rel}^2 cC_d. \tag{6.9}$$

Since we are interested only in the force normal to and tangential to the rotorplane, the lift and drag are projected into these directions (see Figure 6.3):

$$p_N = L\cos\phi + D\sin\phi \tag{6.10}$$

and:

$$p_T = L\sin\phi - D\cos\phi \tag{6.11}$$

The equations (6.10) and (6.11) are normalized with respect to $\frac{1}{2}\rho V_{rel}^2 c$ yielding:

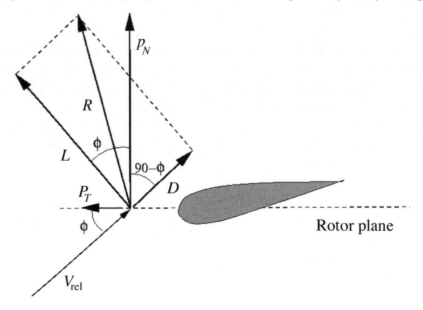

R is the vector sum of the lift L and the drag D; p_N and p_T are the normal and tangential components of R respectively.

Figure 6.3 *The local loads on a blade*

$$C_n = C_l \cos \phi + C_d \sin \phi \tag{6.12}$$

and:

$$C_t = C_l \sin \phi - C_d \cos \phi \tag{6.13}$$

where:

$$C_n = \frac{p_N}{\frac{1}{2} \rho V_{rel}^2 c} \tag{6.14}$$

and:

$$C_t = \frac{p_T}{\frac{1}{2} \rho V_{rel}^2 c} \tag{6.15}$$

From Figure 6.2 it is readily seen from the geometry that:

$$V_{rel} \sin \phi = V_o (1 - a) \tag{6.16}$$

and:

$$V_{rel} \cos \phi = \omega r (1 + a') \tag{6.17}$$

Further, a solidity σ is defined as the fraction of the annular area in the control volume which is covered by blades:

$$\sigma(r) = \frac{c(r)B}{2\pi r} \tag{6.18}$$

B denotes the number of blades, $c(r)$ is the local chord and r is the radial position of the control volume.

Since p_N and p_T are forces per length, the normal force and the torque on the control volume of thickness dr are:

$$dT = B p_N dr \tag{6.19}$$

and:

$$dM = r B p_T dr. \tag{6.20}$$

Using equation (6.14) for p_N and equation (6.16) for V_{rel}, equation (6.19) becomes:

$$dT = \frac{1}{2} \rho B \frac{V_o^2 (1-a)^2}{\sin^2 \phi} cC_n dr. \qquad (6.21)$$

Similarly, if equation (6.15) is used for p_T and equations (6.16) and (6.17) are used for V_{rel}, equation (6.20) becomes:

$$dM = \frac{1}{2} \rho B \frac{V_o (1-a)\omega r (1+a')}{\sin \phi \cos \phi} cC_t r dr. \qquad (6.22)$$

If the two equations (6.21) and (6.4) for dT are equalized and the definition of the solidity equation (6.18) is applied, an expression for the axial induction factor a is obtained:

$$a = \frac{1}{\dfrac{4 \sin^2 \phi}{\sigma C_n} + 1}. \qquad (6.23)$$

If equations (6.22) and (6.5) are equalized, an equation for a' is derived:

$$a' = \frac{1}{\dfrac{4 \sin \phi \cos \phi}{\sigma C_t} - 1}. \qquad (6.24)$$

$= 2 \sin(2\phi)$

Now all necessary equations for the BEM model have been derived and the algorithm can be summarized as the 8 steps below. Since the different control volumes are assumed to be independent, each strip can be treated separately and the solution at one radius can be computed before solving for another radius; in other words for each control volume the following algorithm is applied.

Step (1)   Initialize a and a', typically $a = a' = 0$.
Step (2)   Compute the flow angle ϕ using equation (6.7).
Step (3)   Compute the local angle of attack using equation (6.6).
Step (4)   Read off $C_l(\alpha)$ and $C_d(\alpha)$ from table.
Step (5)   Compute C_n and C_t from equations (6.12) and (6.13).
Step (6)   Calculate a and a' from equations (6.23) and (6.24).
Step (7)   If a and a' has changed more than a certain tolerance, go to step (2) or else finish.
Step (8)   Compute the local loads on the segment of the blades.

This is in principle the BEM method, but in order to get good results it is necessary to apply two corrections to the algorithm. The first is called Prandtl's tip loss factor, which corrects the assumption of an infinite number of blades. The second correction is called the Glauert correction and is an empirical relation between the thrust coefficient C_T and the axial induction factor a for a greater than approximately 0.4, where the relation derived from the one-dimensional momentum theory is no longer valid. Each of these corrections will be treated in separate sections.

After applying the BEM algorithm to all control volumes, the tangential and normal load distribution is known and global parameters such as the mechanical power, thrust and root bending moments can be computed. One has to be careful, however, when integrating the tangential loads to give the shaft torque. The tangential force per length $p_{T,i}$ is known for each segment at radius r_i and a linear variation between r_i and r_{i+1} is assumed (see Figure 6.4). The load p_T between r_i and r_{i+1} is thus:

$$p_T = A_i r + B_i \tag{6.25}$$

where:

$$A_i = \frac{p_{T,i+1} - p_{T,i}}{r_{i+1} - r_i} \tag{6.26}$$

and:

$$B_i = \frac{p_{T,i} r_{i+1} - p_{T,i+1} r_i}{r_{i+1} - r_i} \tag{6.27}$$

The torque dM for an infinitesimal part of the blade of length dr is:

$$dM = r p_T dr = (A_i r^2 + B_i r) dr \tag{6.28}$$

and the contribution $M_{i,i+1}$ to the total shaft torque from the linear tangential load variation between r_i and r_{i+1} is thus:

$$M_{i,i+1} = \left[\frac{1}{3} A_i r^3 + \frac{1}{2} B_i r^2\right]_{r_i}^{r_{i+1}} = \frac{1}{3} A_i (r_{i+1}^3 - r_i^3) + \frac{1}{2} B_i (r_{i+1}^2 - r_i^2). \tag{6.29}$$

The total shaft torque is the sum of all the contributions $M_{i,i+1}$ along one blade multiplied by the number of blades:

$$M_{tot} = B \sum_{1}^{N-1} M_{i,i+1}. \tag{6.30}$$

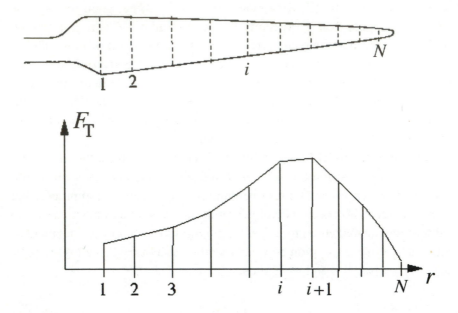

Figure 6.4 *A linear variation of the load is assumed between two different radial positions* r_i *and* r_{i+1}

Prandtl's Tip Loss Factor

As already mentioned, Prandtl's tip loss factor corrects the assumption of an infinite number of blades. For a rotor with a finite number of blades the vortex system in the wake is different from that of a rotor with an infinite number of blades. Prandtl derived a correction factor F to equations (6.4) and (6.5):

$$dT = 4\pi r\rho V_o^2 a(1 - a)Fdr \qquad (6.31)$$

and:

$$dM = 4\pi r^3\rho V_o\omega(1 - a)a'Fdr. \qquad (6.32)$$

F is computed as:

$$F = \frac{2}{\pi}\cos^{-1}(e^{-f}), \qquad (6.33)$$

where:

$$f = \frac{B}{2} \frac{R-r}{r\sin\phi} \qquad (6.34)$$

B is the number of blades, R is the total radius of the rotor, r is the local radius and ϕ is the flow angle. Using equations (6.31) and (6.32) instead of equations (6.4) and (6.5) in deriving the equations for a and a' yields:

$$a = \frac{1}{\dfrac{4F\sin^2\phi}{\sigma C_n} + 1} \qquad (6.35)$$

and:

$$a' = \frac{1}{\dfrac{4F\sin\phi\cos\phi}{\sigma C_t} - 1} \qquad (6.36)$$

Equations (6.35) and (6.36) should be used instead of equations (6.23) and (6.24) in step 6 of the BEM algorithm and an extra step computing Prandtl's tip loss factor F should be put in after step 2. Deriving Prandtl's tip loss factor is very complicated and is not shown here, but a complete description can be found in Glauert (1935).

Glauert Correction for High Values of a

When the axial induction factor becomes larger than approximately 0.4, the simple momentum theory breaks down (see Figure 4.5, where the different states of the rotor are also shown). Different empirical relations between the thrust coefficient C_T and a can be made to fit with measurements, for example:

$$C_T = \begin{cases} 4a(1-a)F & a \le \dfrac{1}{3} \\ 4a(1-\dfrac{1}{4}(5-3a)a)F & a > \dfrac{1}{3} \end{cases} \qquad (6.37)$$

or:

$$C_T = \begin{cases} 4a(1-a)F & a \le a_c \\ 4(a_c^2 + (1-2a_c)a)F & a > a_c \end{cases} \qquad (6.38)$$

The last expression is found in Spera (1994) and a_c is approximately 0.2. F is Prandtl's tip loss factor and corrects the assumption of an infinite number of blades. In Figure 6.5 the two expressions for $C_T(a)$ are plotted for $F = 1$ and compared to the simple momentum theory.

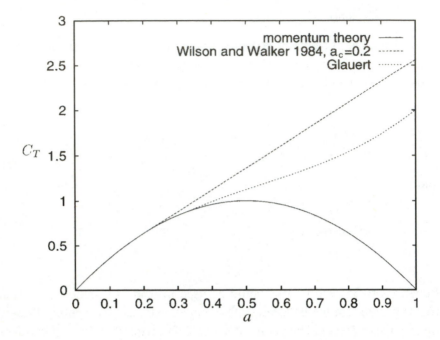

Figure 6.5 *Different expressions for the thrust coefficient C_T versus the axial induction factor* a

From the local aerodynamics the thrust dT on an annular element is given by equation (6.21). For an annular control volume, C_T is by definition:

$$C_T = \frac{dT}{\frac{1}{2}\rho V_o^2\, 2\pi r dr}. \tag{6.39}$$

If equation (6.21) is used for dT, C_T becomes:

$$C_T = \frac{(1-a)^2 \sigma C_n}{\sin^2 \phi} \tag{6.40}$$

This expression for C_T can now be equated with the empirical expression (6.38).

If $a < a_c$:

$$4a(1-a)F = \frac{(1-a)^2 \, \sigma C_n}{\sin^2 \phi} \tag{6.41}$$

and this gives:

$$a = \frac{1}{\dfrac{4F \sin^2 \phi}{\sigma C_n} + 1} \tag{6.42}$$

which is the normal equation (6.35).

If $a > a_c$:

$$4(a^2_c + (1-2a_c)a)F = \frac{(1-a)^2 \sigma C_n}{\sin^2 \phi} \tag{6.43}$$

and this gives:

$$a = \frac{1}{2}\left[2 + K(1-2a_c) - \sqrt{(K(1-2a_c)+2)^2 + 4(Ka^2_c - 1)}\,\right] \tag{6.44}$$

where:

$$K = \frac{4F \sin^2 \phi}{\sigma C_n} \tag{6.45}$$

In order to compute the induced velocities correctly for small wind speeds, equations (6.44) and (6.42) must replace equation (6.35) from the simple momentum theory.

Annual Energy Production

The BEM method has now been derived and it is possible to compute a power curve, in other words the shaft power as a function of the wind speed V_o. In order to compute the annual energy production it is necessary to combine this production curve with a probability density function for the wind. From this function the probability, $f(V_i < V_o < V_{i+1})$, that the wind speed lies between V_i and V_{i+1} can be computed. Multiplying this with the total number of hours per year gives the number of hours per year that the wind speed lies in the interval $V_i < V_o < V_{i+1}$. Multiplying this by the power (in kW) produced by the wind turbine when the wind speed is between V_i and V_{i+1}

gives the contribution of the total production (in kWh) for this interval. The wind speed is discretized into N discrete values (V_i, $i = 1,N$), typically with 1m/s difference (see Figure 6.6).

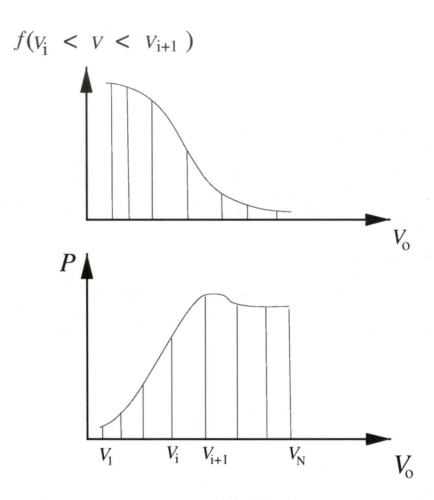

$$f(V_i < V < V_{i+1})$$

Figure 6.6 *Probability* f(V$_i$ < V$_o$ < V$_{i+1}$) *that the wind speed lies between* V$_i$ *and* V$_{i+1}$ *and a power curve in order to compute the annual energy production for a specific turbine on a specific site*

It must be noted that the production must be corrected for losses in the generator and gearbox, which have a combined efficiency of approximately 0.9. Typically the probability density function of the wind is given by either a Rayleigh or a Weibull distribution. The Rayleigh distribution is given by the mean velocity only as:

$$h_R(V_o) = \frac{\pi}{2} \frac{V_o}{\overline{V}^2} \exp\left(-\frac{\pi}{4}\left(\frac{V_o}{\overline{V}}\right)^2\right) \tag{6.46}$$

In the more general Weibull distribution, some corrections for the local siting (for example for landscape, vegetation, and nearby houses and other obstacles) can be modelled through a scaling factor A and a form factor k:

$$h_w(V_o) = \frac{k}{A}\left(\frac{V_o}{A}\right)^{k-1} \exp\left(-\left(\frac{V_o}{A}\right)^k\right) \tag{6.47}$$

The parameters k and A must be determined from local meteorological data, nearby obstacles and landscape. Help in doing this can be obtained from the *European Wind Atlas* (Troen and Petersen,1989). From the Weibull distribution, the probability $f(V_i < V_o < V_{i+1})$ that the wind speed lies between V_i and V_{i+1} is calculated as:

$$f(V_i < V_o < V_{i+1}) = \exp\left(-\left(\frac{V_i}{A}\right)^k\right) - \exp\left(-\left(\frac{V_{i+1}}{A}\right)^k\right) \tag{6.48}$$

The total annual energy production can thus be evaluated as:

$$AEP = \sum_{i=1}^{N-1} \frac{1}{2}(P(V_{i+1}) + P(V_i)) \cdot f(V_i < V_o < V_{i+1}) \cdot 8760. \tag{6.49}$$

Example

Having derived the BEM method and shown how annual energy production can be calculated, it is time for a simple example to illustrate the accuracy that can be obtained for a real turbine. The following example is of a Nordtank NTK 500/41 wind turbine. The turbine is stall regulated (fixed pitch) and the main parameters are listed below:

Rotational speed: 27.1 rpm; $= 2.8 \, rad/s$
Air density: 1.225kg/m³;
Rotor radius: 20.5m;
Number of blades: 3;
Hub height: 35.0m; and
Cut-in wind speed: 4m/s; cut-out wind speed: 25m/s.

Blade description

r [m]	twist [degrees]	chord [m]
4.5	20.0	1.63
5.5	16.3	1.597
6.5	13.0	1.540
7.5	10.05	1.481
8.5	7.45	1.420
9.5	5.85	1.356
10.5	4.85	1.294
11.5	4.00	1.229
12.5	3.15	1.163
13.5	2.60	1.095
14.5	2.02	1.026
15.5	1.36	0.955
16.5	0.77	0.881
17.5	0.33	0.806
18.5	0.14	0.705
19.5	0.05	0.545
20.3	0.02	0.265

Since the power depends directly on the air density ρ, according to the standards the computations must be performed for $\rho = 1.225 \text{kg/m}^3$. The difficult part is to find reliable aerofoil data $C_l(\alpha)$ and $C_d(\alpha)$ for the different aerofoils applied along the span. The data available in the literature are for thin aerofoils not much thicker than 20 per cent of the chord and for angles of attack only slightly above $C_{l,max}$. For structural reasons it is desirable to use very thick aerofoils of approximately 40 per cent of the chord at the root of the blades in order to absorb the high bending moments. Further, the boundary layers on the rotating blades are influenced by centrifugal and Coriolis forces, which alter the post-stall lift and drag coefficients from those measured in a wind tunnel. It is therefore clear that it requires significant engineering skill and experience to construct good aerofoil data for thick aerofoils at high angles of attack including 3-D effects. In Snel et al (1993) and Chaviaropoulos and Hansen (2000) some guidelines are given to correct for the rotational effects. When the power curve from an actual wind turbine is known from measurements, it is common to calibrate the aerofoil data afterwards in order to achieve better agreement between measurements and computations. If a new blade is constructed which is not too different from blades for which the aerofoil data have been calibrated, it is possible to predict the power curve very well. But if a new blade is to be designed with completely new aerofoils, one has to be very careful in using the computed results. The actual geometry and aerofoil data on the blade of the Nordtank NTK 500/41 is not shown here, but an educated guess has been applied to extrapolate the data into high angles of attack. The power curve has been measured (Paulsen, 1995) and the comparison between the computed and measured power curve is shown in Figure 6.7 to give an idea of the accuracy of the BEM model.

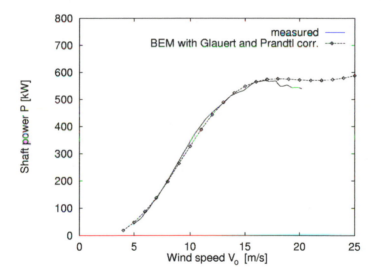

Figure 6.7 *Comparison between computed and measured power curve, i.e. mechanical shaft power as a function of the wind speed*

It is seen in Figure 6.7 that except for at very high wind speeds the BEM method captures the measurements very well. The power curve is often shown in non-dimensional form as in Figure 6.8 or 6.9. Figure 6.8 shows that this particular wind turbine has a maximum efficiency of approximately $C_p = 0.5$ for a tip speed ratio λ between 9 and 10. The advantage of plotting the power coefficient as a function of the inverse tip speed ratio is that λ^{-1} increases linearly with the wind speed V_o.

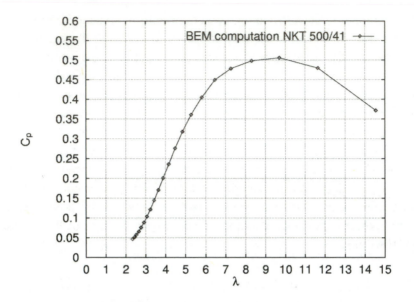

Figure 6.8 *Power coefficient* C_p *as a function of the tip speed ratio* $\lambda = \omega R/V_o$

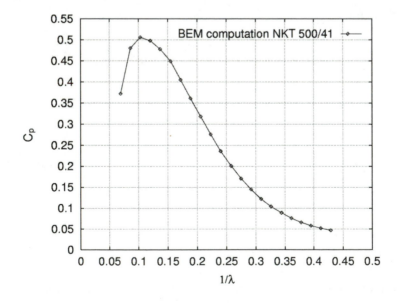

Figure 6.9 *Power coefficient* C_p *as a function of the inverse tip speed ratio* $\lambda^{-1} = V_o/\omega R$

If the turbine were erected at a site where the Weibull parameters are $k = 1.9$ and $A = 6.8$m/s, the annual energy output from the mechanical power would be $1.09 \cdot 10^6$kWh, which corresponds to the consumption of approximately 250 households. The actual number it could supply would be less, however, since the losses from the generator and gearbox have not been taken into account. It should also be remembered that the Weibull parameters vary from site to site and have to be evaluated for each individual siting.

The example shown here is for a stall regulated wind turbine, but the BEM method can also be used to predict the necessary pitch setting of a pitch regulated wind turbine. When the pitch is referred to for an entire wind turbine, it means the angle between the chord line of the tip aerofoil and the rotor plane. A pitch regulated wind turbine may operate at a fixed pitch until a certain nominal power is generated. For higher wind speeds the blades are pitched normally with the leading edge into the wind in order to keep this nominal power. Therefore the power curve of a pitch regulated wind turbine is absolutely flat after the nominal power has been reached. More runs are required with the BEM method to predict the power curve and the pitch setting for different wind speeds. One procedure for computing a pitch regulated wind turbine is sketched in Figure 6.10.

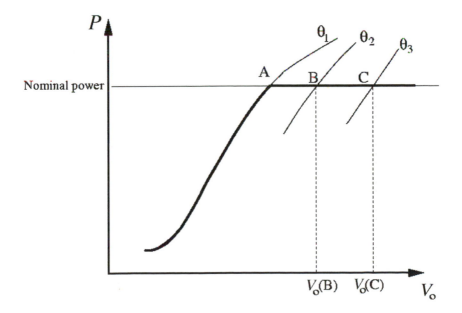

Figure 6.10 *Sketch of a power curve for a pitch controlled wind turbine*

At point A the nominal power is reached and it is necessary to change the pitch. BEM calculations are made for a pitch of θ_2 and θ_3. It is seen that for a wind speed of $V_o(B)$ the pitch much be changed to θ_2 in order to obtain the nominal power. The BEM method as derived in this chapter is steady, so it is not possible to compute the transient from point A to point B and from point B to point C. This requires an extended BEM method as shown in a later chapter.

References

Chaviaropoulos, P. K. and Hansen, M. O. L. (2000) 'Investigating three-dimensional and rotational effects on wind turbine blades by means of a quasi-3D Navier-Stokes solver', *Journal of Fluids Engineering*, vol 122, pp330–336

Glauert, H. (1935) 'Airplane propellers', in W. F. Durand (ed) *Aerodynamic Theory*, vol 4, Division L, Julius Springer, Berlin, pp169–360

Paulsen, U. S. (1995) 'Konceptundersøgelse Nordtank NTK 500/41 Strukturelle Laster' (in Danish), Risø-I-936(DA), Risø National Laboratory

Snel, H., Houwink, B., Bosschers, J., Piers, W. J., van Bussel, G. J. W. and Bruining, A. (1993) 'Sectional prediction of 3-D effects for stalled flow on rotating blades and comparison with measurements', in *Proceedings of European Community Wind Energy Conference 1993*, pp395–399

Spera, D. A. (1994) *Wind Turbine Technology*, ASME Press, New York

Troen, I. and Petersen, E. L. (1989) *European Wind Atlas*, Risø National Laboratory

7

Control/Regulation
and Safety Systems

The control or regulation system ensures that the turbine operates within the design range, in other words that it:

- keeps the rotational speed within a certain range;

- yaws the turbine;

- keeps the power output within a certain range; or

- starts and stops the turbine.

Further, the control system can ensure a smooth power output $P(t)$ and/or may optimize the power output at lower wind speeds. To limit the power at high wind speeds, the following three strategies may be used, the first two being by far the most common:

1 stall regulation;

2 pitch regulation; and

3 yaw control.

Stall Regulation

Stall regulation is mechanically the most simple, since the blades are fixed to the hub and cannot be pitched. A stall regulated wind turbine is normally operated at an almost constant rotational speed and thus the angle of attack increases as the wind speed increases (see Figure 6.2). Eventually, as the local angles of attack are increased, the blades stall, causing the lift coefficient to decrease and the drag coefficient to increase, yielding a lower tangential load according to equation (6.13). The power decrease depends on the pitch angle (angle between rotor plane and tip chord), the twist and chord distributions, and the aerofoils used for the blades. If a site test shows that the power is not limited sufficiently, it is necessary to unbolt the blades and change their fixed pitch setting. On a stall regulated wind turbine an asynchronous generator is often used whereby the rotational speed is almost constant and determined by the torque characteristic of the generator, in other words the shaft torque into the generator, M_G, as a function of the rotational speed of this shaft, n. A typical torque characteristic is illustrated in Figure 7.1, where it is seen that the asynchronous generator can act both as a motor and as a generator. The motor mode can be used to start the turbine. The sign here is deemed positive when the generator is producing electricity. The rotational speed of the generator will be between n_o and n_{nom} and the torque will equal the torque produced by the rotor blades M_R at the generator shaft. The rotational speed of the generator for zero shaft torque, n_o, for an asynchronous generator is:

$$n_o = 60 f_{grid}/p, \qquad (7.1)$$

where f_{grid} is the frequency of the grid (in Europe 50Hz) and p denotes the number of pole pairs. n_o is thus 1500 rpm for 4 poles and 1000 rpm for 6 poles. The rotational speed of the generator is higher than the rotational speed of the rotor and therefore there is a gearbox between the generator and the rotor. The relationship between the rotational speed of the rotor, ω, and the rotational speed of the generator, n, is given through the transmission factor r as $\omega = n/r$. The relative difference between the actual rotational speed n and n_o is called the slip $SL = (n - n_o)/n_o$ and for a normal stall regulated wind turbine the slip is about 1–3 per cent. This means that the rotational speed of the rotor is almost constant and the possibility of using the rotor as a flywheel to store energy – from a gust, for example – is small. Changes in the rotor torque M_R from, for example, turbulence in the wind are thus almost immediately transferred to the generator torque M_G and thus to the produced electrical power $P_{EL} = M_G 2\pi n/60$. Consider a wind turbine operating at point A on Figure 7.1 and the wind speed increasing. In this case the torque, M_R, from the rotor blades will also increase and the rotor accelerate according to equation (7.2):

$$I\frac{d\omega}{dt} = M_R - M_G, \qquad (7.2)$$

until M_G again equals M_R at point B. I denotes the moment of inertia of the rotor about the rotational axis and here M_R and M_G are the rotor and generator torque at the rotor shaft.

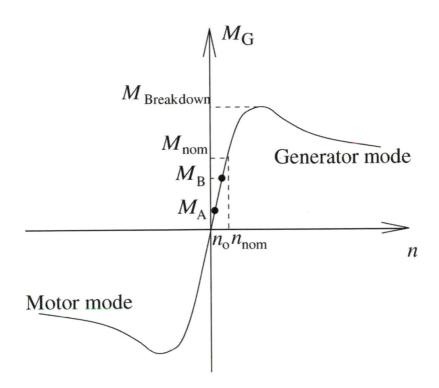

Figure 7.1 *A typical torque characteristic for an asynchronous generator*

If M_R exceeds the maximum point on the torque characteristic, $M_{Break\ down}$, or the generator is disconnected from the grid, the term $M_R - M_G$ on the right–hand side of equation (7.2) is always positive and the rotor will start to accelerate; in this case the rotational speed can become so high that a risk of breakdown exists. The safety system must detect this and ensure that the rotor is stopped. On a stall regulated wind turbine it is common that the outer parts of the blades are activated by centrifugal force to turn 90° and thus act as an aerodynamic brake limiting M_R, (see Figure 7.2).

An example of a time history of a start-up at a high wind speed for a stall regulated wind turbine is shown in Figure 7.3. This figure shows actual measurements performed on the Elkraft 1MW demonstration wind turbine, sited near Copenhagen. This turbine can run as a stall regulated machine as well as a pitch regulated machine (see later). The first curve in Figure 7.3 shows the wind speed at hub height, the second curve shows the rotational speed of the generator n and the last curve shows the corresponding power as a function of time. When starting the turbine, at $t = 420$s, the generator is switched off; in other words M_G is zero and the rotor starts to accelerate until

$n = n_o$ at $t = 445$s. Then the generator is connected instantaneously, giving rise to an overshoot in the power, as seen directly in the power time history.

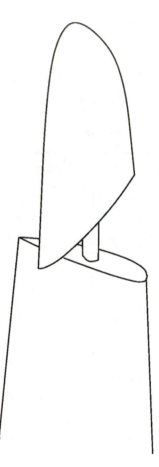

Figure 7.2 *Turnable tip used as an aerodynamic brake and activated by the centrifugal force*

Pitch Regulation (Constant Rotational Speed)

On a pitch regulated machine it is possible to actively pitch the entire blade and thus to change simultaneously the angles of attack along its entire length. One way of controlling the pitch is sketched in Figure 7.4, where a piston placed within the main shaft changes the pitch. The position of the piston is determined by applied hydraulic pressure. If the hydraulic pressure is lost a

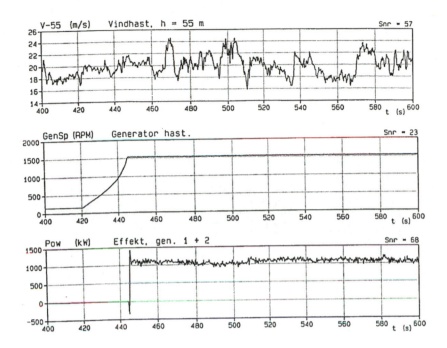

Figure 7.3 *Starting a stall regulated wind turbine at high wind speed: Elkraft 1MW demonstration wind turbine at Avedøre, Denmark*

spring will retract the piston, twisting the leading edges of the blades up against the wind. It is worth mentioning here that this is not the only way to pitch the blades: each blade could be fitted with a small electrical motor, meaning that they could be pitched independently.

A pitched blade can act as an aerodynamic brake and it is no longer necessary to include tip brakes as on a stall regulated machine. By pitching the entire blade it is possible to control the angles of attack and thus the power output. Normally, the power is reduced by decreasing the angles of attack by pitching the leading edge of the blades up against the wind, in other words by increasing θ in the expression for the angle of attack $\alpha = \phi - \theta$, where ϕ is the flow angle. Alternatively, one could reduce the power output by increasing the angle of attack, thus forcing the blades to stall. This is called active stall. Due to the turbulent nature of the wind, the instantaneous power output of a pitch regulated machine will often exceed the rated power, the timescales of these fluctuations being smaller than the time it takes to pitch the blades (see Figure 7.5). Figure 7.5 shows a start-up of the same

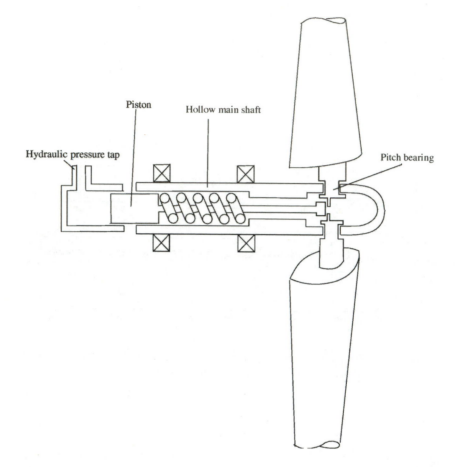

Figure 7.4 *Sketch of mechanism to change the pitch of the blades through a piston placed inside the main shaft*

turbine as in Figure 7.3 but now running as a pitch controlled machine. It is seen that the start-up occurs much more smoothly since the blades are gradually pitched from 50° to about 15° (second curve in Figure 7.5). Comparing the time history of the power output after the start-up for the turbine running as a pitch regulated wind turbine (Figure 7.5) with the time history of the power output for the same turbine running as a stall regulated wind turbine (Figure 7.3), it is seen that the peaks are smaller when the turbine is running as a stall regulated turbine. Since turbulent fluctuations occur much faster than the time it takes to pitch the blades, the power from a pitch regulated wind turbine follows for some time the stationary power curve for a fixed pitch turbine (see Figure 6.10). For high wind speeds the stationary power curve for a pitch regulated machine at a fixed pitch has a much higher slope dP/dV_o than the corresponding stationary power curve for a stall regulated wind turbine and thus a larger variation $\Delta P = dP/dV_o \cdot \Delta V_o$ for the wind speed interval ΔV_o. This is the reason why the power fluctuations are lower for a stall regulated wind turbine than for a pitch regulated machine at high wind speeds. The control diagram when the pitch is used to control the power is shown in Figure 7.6. The controller of a classical pitch regulated wind turbine is not, however, responding to the wind speed but reacts directly to the power as:

$$\frac{d\theta_p}{dt} = \frac{KI(P - P_{ref})}{1 + \dfrac{\theta_p(t)}{KK}}, \tag{7.3}$$

where KI is an integration constant and KK is the gain reduction that reduces the pitch rate at high values of the pitch angle itself. High values of the pitch angle correspond to high wind speeds, where the loads are also very big and thus very sensitive to the pitch angle.

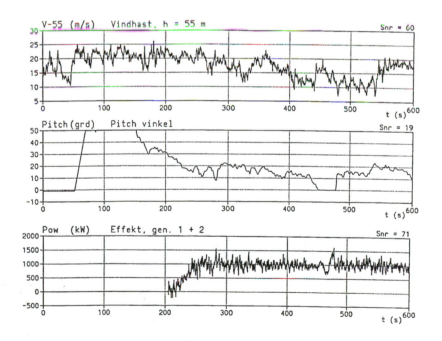

Figure 7.5 *Starting a pitch regulated wind turbine at high wind speed: Elkraft 1MW demonstration wind turbine at Avedøre, Denmark*

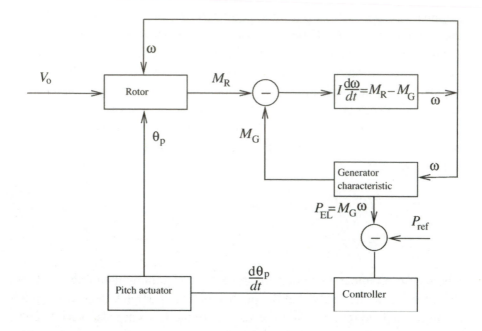

Figure 7.6 *Control diagram for a pitch regulated wind turbine controlling the power*

To overcome the problem of the large peaks in power and loads on a pitch regulated wind turbine at high wind speeds, a system called OptiSlip® is used by VESTAS. OptiSlip® utilizes the fact that the torque characteristic; in other words the slip, for an asynchronous generator can be altered by changing an inner resistance R in the generator. A torque characteristic with a constant torque, after a certain value, $M_G = M_{nom} =$ const, can thus be obtained as indicated by the thick curve in Figure 7.7, which shows the effective characteristic obtained with different resistances, $R_1, R_2, R_3, \ldots$. In high winds the power output from a wind turbine using OptiSlip® will be almost constant, with very small fluctuations around the nominal value.

When the rotor torque M_R exceeds M_{nom} the rotor will start to accelerate in accordance with equation (7.2). The control system detects this and starts pitching the blades. The advantage of this is that the time needed to accelerate the rotor is longer than that needed to pitch the blades, so there is enough time available to physically move the blades, and at the same time a much smoother power output is obtained since the torque on the generator shaft is almost constant. Further, the loads are also reduced, which increases the fatigue lifetime. The pitch system is thus controlling the rotational speed and not the power.

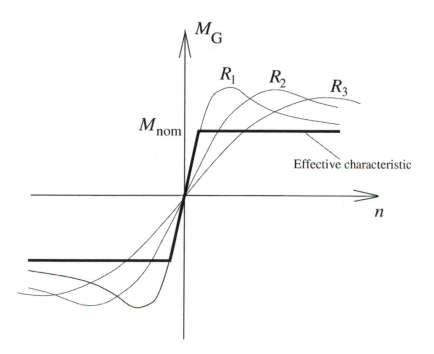

Figure 7.7 *Torque characteristic for an asynchronous generator with variable slip*

There follows an example illustrating stall regulation, pitch regulation and active stall, based on the previously introduced stall controlled NTK 500/41 wind turbine. Using the same blades, the same rotational speed and the same Weibull parameters, the potential increase from a pitch mechanism in the annually captured energy is investigated. It is decided to use the same generator; in other words the nominal mechanical power remains unaltered at approximately 580kW. For each wind speed an optimum pitch angle is sought by varying the pitch in a BEM calculation (see Figure 7.8). It is seen that the mechanical power for a constant wind speed has an optimum value for the global pitch, but that for this turbine the variation is small.

For higher wind speeds the optimum value of the power exceeds the nominal power, as shown in Figure 7.9. In this figure it is also seen that there exist two choices of pitch yielding the nominal power. The smaller value, θ_p = 0.16°, corresponds to active stall, since the local angles of attack are higher than the limit for unseparated flow. The higher value, θ_p = 20.2°, corresponds to classical pitch regulation, whereby the local angles of attack and thus the loads on the blades are reduced.

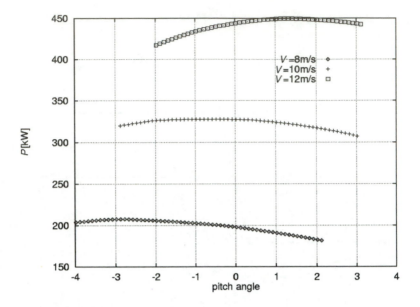

Figure 7.8 *Optimum pitch angle for different wind speeds*

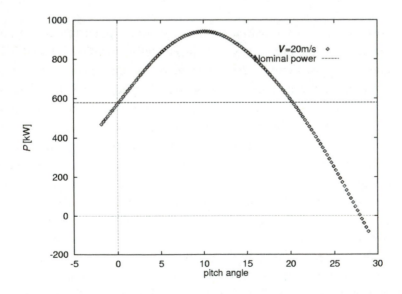

Figure 7.9 *Variation of the mechanical power with the pitch for a wind speed of 20m/s on an NTK 500/41 wind turbine*

The computed mechanical power curves for the NTK 500/41 wind turbine running as a stall and pitch controlled wind turbine are plotted in Figure 7.10. For the variable pitch machine, the optimum values for the pitch have been used for the lower wind speeds. These optimum pitch values are shown in Figure 7.11 as a function of the wind speed. For wind speeds slightly higher than 14m/s, the blades must be pitched to ensure a power below the nominal value. The lower branch in Figure 7.11 shows the pitch setting on a machine controlled by active stall and the upper branch shows the pitch setting on a pitch controlled wind turbine.

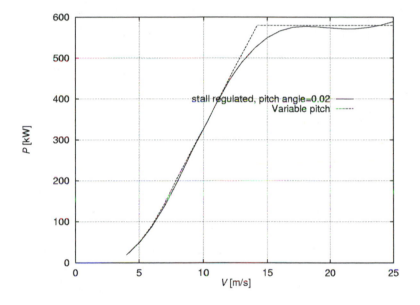

Figure 7.10 *Computed power curve for the NTK 500/41 wind turbine running as a stall controlled or pitch regulated machine*

Assuming the same Weibull parameters as in the example concerning the NTK 500/41 machine, i.e. $k = 1.9$ and $A = 6.8$m/s, the annual energy production from the turbine running as a pitch regulated machine is $1.11 \cdot 10^6$ kWh. In this example an increase in the annual energy production of approximately 2 per cent has thus been achieved by changing from a stall to a pitch regulated machine. (It should be noted that the annual energy production in both cases has been computed using mechanical power; in other words the losses in the generator and in the gearbox have not been taken into account.) The main contribution to this increase comes from the

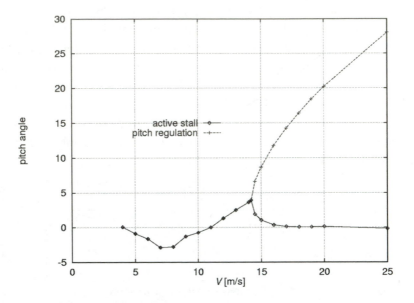

Figure 7.11 *Optimum values for the pitch and the necessary pitch to avoid exceeding the nominal power*

shape of the power curve just before reaching the nominal power, where the power curve of the pitch regulated wind turbine is steeper than that of the stall regulated machine. Due to the turbulent characteristics of the wind, however, the power curve on a real pitch regulated machine is not so steep close to P_{nom} as the theoretical curve indicates in Figure 7.10, so the increase in the annual production is probably slightly lower than estimated from the theoretical power curves.

Yaw Control

Instead of limiting the power output using pitch or stall regulation, it is possible to control the yaw of the turbine. On normal pitch or stall regulated machines, it is common to have a yaw drive, which is constantly trying to rotate the nacelle to minimize the yaw misalignment in order to get as much air through the rotor disc as possible. On a yaw controlled machine, the rotor is turned out of the wind in high winds to limit the airflow through the rotor and thus the power extraction. Yaw control was used on the old Western mills. For larger machines, only the 1.5MW Italian prototype called GAMMA 60 is, to the author's knowledge, yaw controlled. Therefore yaw control will not be treated further in this text.

Variable Speed

The power coefficient is a function of the tip speed ratio and the pitch angle, $C_p(\theta_p, \lambda)$, and by applying variable speed on a pitch regulated rotor it is possible to run the turbine at the optimum point $C_{p,\,max}$ occurring at $\theta_{p,opt}$ and λ_{opt}. From Figure 6.8, which shows the C_p-λ curve for the NTK 500/41 machine, it is seen that the turbine for this pitch angle runs most efficiently at λ approximately equal to 9.8. From a C_p-λ curve, a plot showing the power P as a function of the rotational speed ω for different wind speeds can be derived as:

$$P = \frac{1}{2}\rho V_o^3 A C_p(\lambda) \tag{7.4}$$

and:

$$\omega = \frac{\lambda V_o}{R}. \tag{7.5}$$

This has been done for the NTK 500/41 machine and the result is shown in Figure 7.12. The turbine is equipped with an asynchronous generator forcing the blades to rotate at 27.1 rpm, indicated by the vertical line. It is seen that the turbine is running most efficiently at a wind speed V_o of approximately 7m/s. Some stall and pitch regulated wind turbines using asynchronous generators, in other words running at a fixed rotational speed, therefore have two generators, one which is efficient at lower wind speeds and one which is efficient at higher wind speeds. If another type of generator had been used, one which is able to run at different rotational speeds, the turbine could be operated at the optimum rotational speed for each wind speed, as indicated in Figure 7.12 by the operational line that intersects all the top points in the curves for the different wind speeds. All points on this line correspond to the highest C_p that can be obtained for the applied pitch angle. It is noted that at a fixed value of the tip speed ratio λ, the angular velocity ω and thus the tip speed will increase proportionally with the wind speed according to equation (7.5). Due to noise emission, the tip speed is limited to approximately $\omega R = 70$m/s and therefore the optimum tip speed ratio can only be obtained for lower wind speeds. The alternating current produced by such a generator will have a frequency different from the frequency of the grid (50Hz in Europe). Therefore the alternating current (AC) is first transformed into direct current (DC) and then back to alternating current with the correct frequency using a so-called ACDC/DCAC device.

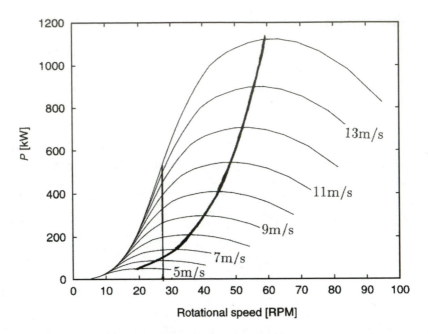

Figure 7.12 *Constant speed versus variable speed*

Combining the fact that power is the torque multiplied by the angular velocity, $P = M\omega$, with equations (7.4) and (7.5) yields:

$$M = \frac{1}{2}\rho\omega^2 R^3 A C_p(\lambda,\theta_p)/\lambda^3. \tag{7.6}$$

The generator characteristic for the highest obtainable power coefficient $C_{p,max}(\theta_{p,opt}, \lambda_{opt})$ is thus:

$$M_{opt} = \frac{1}{2}\rho\omega^2 R^3 A C_{p,max}(\theta_{p,opt},\lambda_{opt})/\lambda^3_{opt} = const \cdot \omega^2. \tag{7.7}$$

This characteristic should be used until the maximum allowable tip speed is reached. Thereafter the torque should be constant, such as the characteristic shown in Figure 7.7. If the rotor torque exceeds this maximum generator torque it will accelerate according to equation (7.2), but since the inertia of the rotor is large it will take some time for the rotor to build up speed. The electrical output will be almost constant since the generator torque is constant, and the pitch system should control the rotational speed and not the power. Since the change in rotational speed is slow due to the high inertia of the rotor, there is plenty of time for the pitch mechanism to act. This will overcome the problem of the large spikes/fluctuations from a pitch regulated wind turbine operating at constant rotational speed as seen in the power output shown in Figure 7.5. The control diagram for a pitch regulated variable speed machine is shown in Figure 7.13.

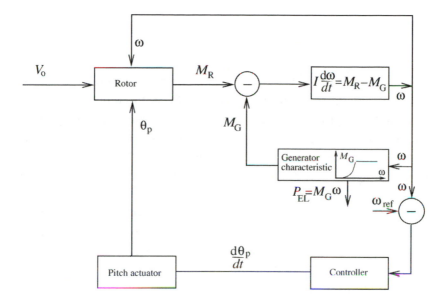

Figure 7.13 *Schematic drawing of a control diagram for a pitch regulated variable speed machine*

One way of obtaining variable speed with an asynchronous generator is to apply a so-called double feed induction generator (DFIG). Here a control device is connected to the generator that effectively controls the net frequency seen by the rotor in the generator. With a DFIG it is thus possible to create a variable speed that can reduce fatigue loads and improve power quality and that for lower wind speeds can produce slightly more power by running at the optimum C_p. However, the extra cost of the control system and the necessary converters also has to be considered.

8

Optimization

Having derived all the necessary equations to compute a given wind turbine, one should be in a position to use these to compute an optimum design. First, an optimum design must be defined, but since the purpose of a wind turbine is to produce electricity, and this should be at a competitive cost, the object function is a design that can last for a typical design lifetime of 20 years and at a given site minimize energy production cost (in $/kWh).

To do this it is necessary to estimate a cost function for every component of the wind turbine and a maintenance cost. From a purely technical point of view, the optimum design could be a wind turbine which for a given rotor diameter captures as many kWh/year as possible. If the turbine is sited where the wind speed V_o is constant in time, it is obvious to optimize the power coefficient at this wind speed. Since in practice wind speed is not constant, however, an optimum design could have lower $C_{p,max}$ as sketched in the second design in Figure 8.1.

As shown in the previous chapter, the annual energy production is a combination of the wind distribution and the power curve. Thus the optimum design also depends on the actual siting. A BEM code as described earlier can be coupled with an optimization algorithm with appropriate constraints to optimize the geometry of the blades, for example. Of course, it is imperative afterwards to verify that the calculated optimum design will also survive the entire design period, taking both extreme and fatigue loads into account. It is possible to compute analytically the geometry of design 1 in Figure 8.1 with the already derived equations as described in, for example, Glauert (1935). Such a design is more interesting for an aeroplane propeller, which is mainly operating at cruise speed, but it could perhaps also be interesting for a wind turbine with variable rotational speed. Ideally such a machine can be kept by a control mechanism at the optimum tip speed ratio, λ_{opt}, and pitch angle, $\theta_{p,opt}$, as described in Chapter 7.

First a good aerofoil is chosen; this must be relatively roughness insensitive and possess an acceptable stall characteristic. Noise considerations might

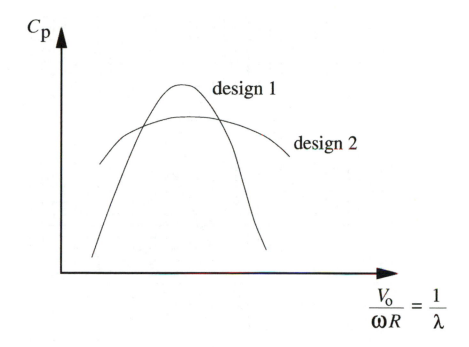

Figure 8.1 *Two different designs: design 1 has a high* $C_{p,max}$ *but* C_p *drops off quickly at different tip speed ratios; design 2 has a lower* $C_{p,max}$ *but performs better over a range of tip speed ratios*

also influence the choice of aerofoil. A possible choice is the NACA63-415 aerofoil, which has proven to possess good stall characteristics on stall regulated wind turbines. Since it is designed to operate at one point only, it is ensured that the effective angle of attack has an optimum value along the entire span. The optimum value is where the ratio between the lift and the drag is highest.

In Figure 8.2 it is seen that the NACA63-415 aerofoil at Re $= 3 \cdot 10^6$ has a maximum value of C_l/C_d of approximately 120 at an angle of attack α of 4°. It is also seen that this maximum value drops to approximately 67 when standard roughness is added to the aerofoil at the same optimum angle of attack of 4°. In the following example the values with roughness are used: $C_{l,opt}(4°) = 0.8$ and $C_{d,opt}(4°) = 0.012$. Further, the number of blades B is chosen as 3 and the design point as $\lambda = \omega R/V_o = 6$. Since an angle of attack of 4° is chosen in the design point, the flow is attached to the blades and equations (4.32) and (4.38) are valid and can be combined to give an optimum

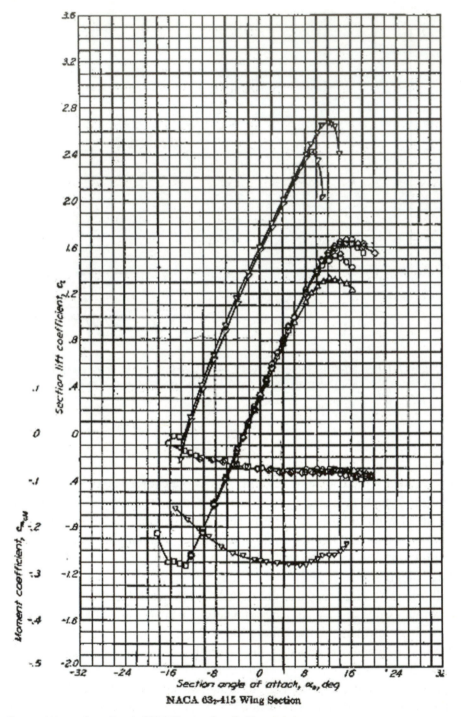

NACA 63₂-415 Wing Section

Source: Abbot and von Doenhoff (1959), reproduced with permission.

Figure 8.2 *Airfoil data for the NACA63-415 airfoil*

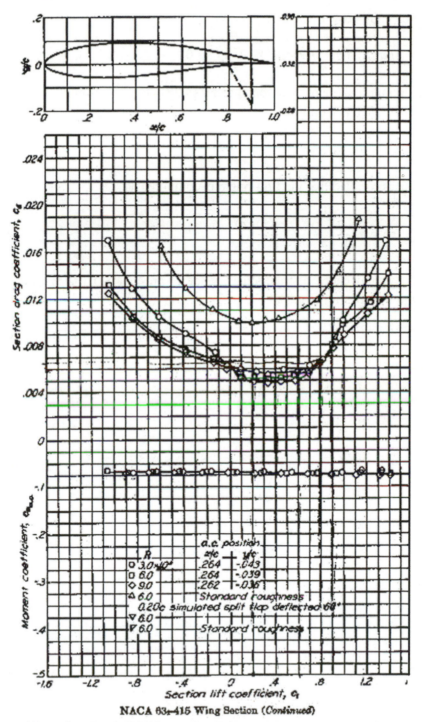

NACA 63₁-415 Wing Section (Continued)

Source: Abbot and von Doenhoff (1959), reproduced with permission.

Figure 8.2 *continued*

relationship between x and a:

$$16a^3 - 24a^2 + a(9 - 3x^2) - 1 + x^2 = 0 \qquad (8.1)$$

The optimum value of a' is found using (4.38) and the optimum local pitch angle can then be computed as:

$$\theta_{opt} = \phi - \alpha_{opt}, \qquad (8.2)$$

since it is recalled that the flow angle is computed as:

$$\tan \phi = \frac{(1-a)V_o}{(1+a')\omega r} = \frac{(1-a)}{(1+a')x}. \qquad (8.3)$$

The optimum chord distribution is found from equation (6.23) using the optimum values for a and a':

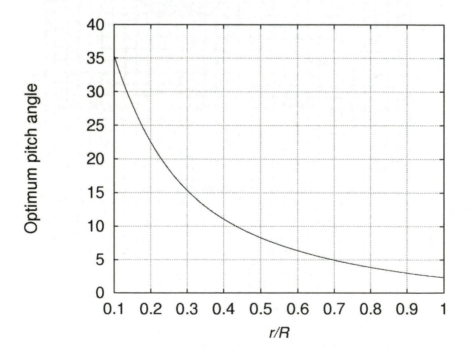

Figure 8.3 *Optimum pitch distribution (neglecting Prandtl's tip loss factor) for $\lambda = 6$, $\alpha_{opt} = 4$, $C_{l,opt} = 0.8$, $C_{d,opt} = 0.012$ and $B = 3$*

$$\frac{c(x)}{R} = \frac{8\pi ax \sin^2\phi}{(1-a)BC_n\lambda},$$

(8.4)

where:

$$C_n = C_{l,opt} \cos(\phi) + C_{d,opt} \sin(\phi)$$

(8.5)

For $\lambda = 6$, $\alpha_{opt} = 4$, $C_{l,opt} = 0.8$, $C_{d,opt} = 0.012$ and number of blades $B = 3$, the optimum chord and pitch distribution can now be computed; the solution is shown graphically in Figures 8.3 and 8.4. Note, however, that in these simple expressions Prandtl's tip loss factor has not been taken into account. Taking Prandtl's tip loss factor into account will change the optimum pitch and chord distribution close to the tip.

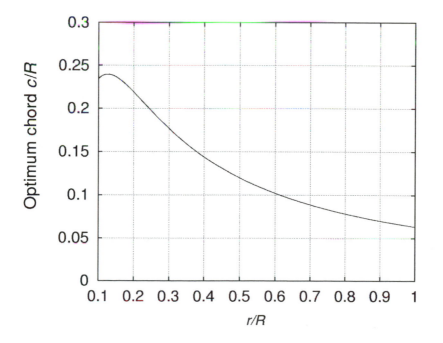

Figure 8.4 *Optimum chord distribution (neglecting Prandtl's tip loss factor) for* $\lambda = 6$, $\alpha_{opt} = 4$, $C_{l,opt} = 0.8$, $C_{d,opt} = 0.012$ *and* $B = 3$

References

Abbot, H. and von Doenhoff, A. E. (1959) *Theory of Wing Sections*, Dover Publications, New York

Glauert, H. (1935) 'Airplane propellers', in W. F. Durand (ed) *Aerodynamic Theory*, vol 4, Division L, Julius Springer, Berlin, pp169–360

9

Unsteady BEM Model

To estimate the annual energy production from a wind turbine at a given site with known wind distribution it is sufficient to apply a steady BEM method as described in Chapter 6 to compute the steady power curve. But due to the unsteadiness of the wind seen by the rotor caused by atmospheric turbulence, wind shear and the presence of the tower it is necessary to use an unsteady BEM method to compute realistically the aeroelastic behaviour of the wind turbine. To do this a complete structural model of the wind turbine is also required; this must be coupled with the unsteady BEM method since, among other things, the velocity of the vibrating blades and the tower change the apparent wind seen by the blades and thus also the aerodynamic loads.

Since the wind changes in time and space it is important at any time to know the position relative to a fixed coordinate system of any section along a blade. The fixed or inertial coordinate system can be placed at the bottom of the tower. Depending on the complexity of the structural model a number of additional coordinate systems can be placed in the wind turbine. The following example illustrates a very simple model, with the wind turbine described by four coordinate systems as shown in Figure 9.1.

First, an inertial system (coordinate system 1) is placed at the base of the tower. System 2 is non-rotating and placed in the nacelle, system 3 is fixed to the rotating shaft and system 4 is aligned with one of the blades. Note that due to the orientation of coordinate system 2, the tilt angle θ_{tilt} must be negative if the shaft is to be nose up as sketched in Figure 9.1.

A vector in one coordinate system $\mathbf{X}_A = (x_A, y_A, z_A)$ can be expressed in another coordinate system $\mathbf{X}_B = (x_B, y_B, z_B)$ through a transformation matrix $\mathbf{a}_{AB}$:

$$\mathbf{X}_B = \mathbf{a}_{AB}\mathbf{X}_A. \tag{9.1}$$

The columns in the transformation matrix $\mathbf{a}_{AB}$ express the unit vectors of system A in system B. Further, the transformation from system B to system A can be found as $\mathbf{a}_{BA} = \mathbf{a}_{AB}^T$. The rules will now be applied to the coordinate systems shown in Figure 9.1. First the transformation matrix $\mathbf{a}_{12}$ will be constructed. To begin with, system 1 and 2 are identical with the exception

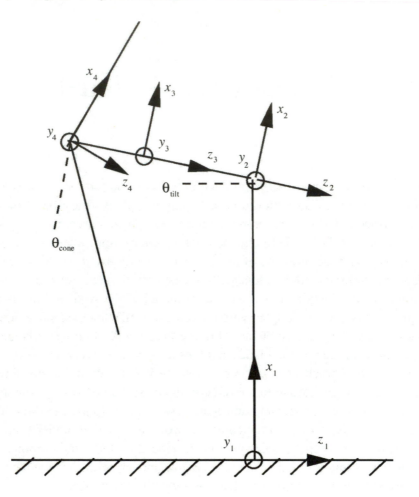

Figure 9.1 *Wind turbine described by four coordinate systems*

of the position of the origins. System 2 is first rotated about the x-axis with the angle θ_{yaw}. This gives a transformation matrix:

$$
\mathbf{a_1} = \begin{pmatrix} 1 & 0 & 0 \\ 0 & \cos\theta_{yaw} & \sin\theta_{yaw} \\ 0 & -\sin\theta_{yaw} & \cos\theta_{yaw} \end{pmatrix}. \tag{9.2}
$$

Hereafter system 2 is rotated θ_{tilt} about the y-axis, yielding a transformation matrix:

$$\mathbf{a_2} = \begin{pmatrix} \cos\theta_{tilt} & 0 & -\sin\theta_{tilt} \\ 0 & 1 & 0 \\ \sin\theta_{tilt} & 0 & \cos\theta_{tilt} \end{pmatrix}. \tag{9.3}$$

Since system 2 is not rotated about the z-axis, this transformation matrix then becomes:

$$\mathbf{a_3} = \begin{pmatrix} 1 & 0 & 0 \\ 0 & 1 & 0 \\ 0 & 0 & 1 \end{pmatrix}. \tag{9.4}$$

The total transformation matrix, $\mathbf{X_2} = \mathbf{a_{12}}\mathbf{X_1}$, between system 1 and system 2 is found as $\mathbf{a_{12}} = \mathbf{a_3}.\mathbf{a_2}.\mathbf{a_1}$.

Since the shaft in this simple model is assumed to be stiff, the only transformation between system 2 and system 3 is a rotation about the z-axis:

$$\mathbf{a_{23}} = \begin{pmatrix} \cos\theta_{wing} & \sin\theta_{wing} & 0 \\ -\sin\theta_{wing} & \cos\theta_{wing} & 0 \\ 0 & 0 & 1 \end{pmatrix}. \tag{9.5}$$

θ_{wing} is the rotation of blade 1 as shown in Figure 9.2.

System 4 is only rotated θ_{cone} about the y-axis and the transformation matrix is thus:

$$\mathbf{a_{34}} = \begin{pmatrix} \cos\theta_{cone} & 0 & -\sin\theta_{cone} \\ 0 & 1 & 0 \\ \sin\theta_{cone} & 0 & \cos\theta_{cone} \end{pmatrix}. \tag{9.6}$$

Note that in order for the blades to cone as shown in Figure 9.1 the cone angle must be negative. A point along blade 1 is described in coordinate system 4 as $\mathbf{r_4} = (x,0,0)$, where x is the radial distance from the rotational axis to the point on the blade. To transform this vector from system 4 to the inertial system 1, the following transformations are thus required:

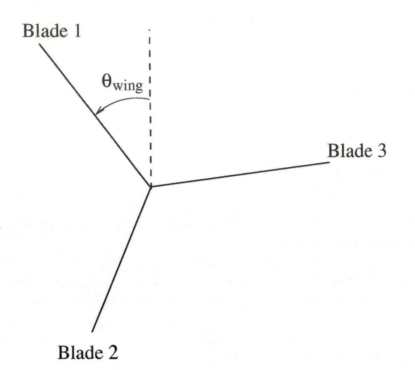

Figure 9.2 *Rotor seen from downstream*

First, the vector is transformed to system 3, which again can be transformed to system 2 and finally to system 1 as:

$$\mathbf{r}_3 = \mathbf{a}_{43}.\,\mathbf{r}_4$$
$$\mathbf{r}_2 = \mathbf{a}_{32}.\,\mathbf{r}_3$$
$$\mathbf{r}_1 = \mathbf{a}_{21}.\,\mathbf{r}_2$$

or directly as:

$$\mathbf{r}_1 = \mathbf{a}_{21}.\,\mathbf{a}_{32}.\,\mathbf{a}_{43}.\mathbf{r}_4 = \mathbf{a}_{12}^{\mathrm{T}}.\,\mathbf{a}_{23}^{\mathrm{T}}.\,\mathbf{a}_{34}^{\mathrm{T}}.\mathbf{r}_4$$

To find the coordinates of the point on the blade in system 1 the vector addition shown in Figure 9.3 is applied:

$$\mathbf{r} = \begin{pmatrix} x_p \\ y_p \\ z_p \end{pmatrix} = \mathbf{r}_t + \mathbf{r}_s + \mathbf{r}_b. \tag{9.7}$$

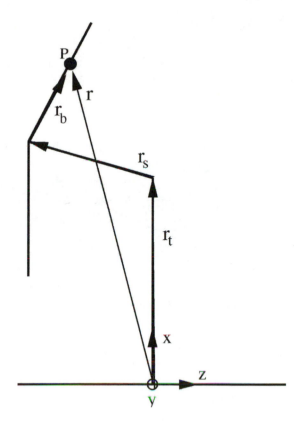

Figure 9.3 *A point on the wing described by vectors*

All the vectors in equation (9.7), $\mathbf{r}_t$, $\mathbf{r}_s$ and $\mathbf{r}_b$, must be given in the same coordinate system. For a position on a blade the natural coordinate system is 1 because the wind velocity is given in the fixed system.

The undisturbed wind velocity seen by the blade is found by transforming this velocity $\mathbf{V}_1$ to system 4:

$$\mathbf{V}_o = \begin{pmatrix} V_x \\ V_y \\ V_z \end{pmatrix} = \mathbf{a}_{34} \cdot \mathbf{a}_{23} \cdot \mathbf{a}_{12} \mathbf{V}_1 = \mathbf{a}_{14} \mathbf{V}_1. \tag{9.8}$$

To find the relative velocity seen by the blade, $\mathbf{V}_{rel}$, the rotational velocity, $\mathbf{V}_{rot}$, plus the induced velocity, $\mathbf{W}$, must be added as vectors to $\mathbf{V}_o$ in system 4 as shown in Figure 9.4.

$$\mathbf{V}_{rel} = \mathbf{V}_o + \mathbf{V}_{rot} + \mathbf{W} \Rightarrow$$

$$\begin{pmatrix} V_{rel,y} \\ V_{rel,z} \end{pmatrix} = \begin{pmatrix} V_y \\ V_z \end{pmatrix} + \begin{pmatrix} -\omega x \cos\theta_{cone} \\ 0 \end{pmatrix} + \begin{pmatrix} W_y \\ W_z \end{pmatrix} \qquad (9.9)$$

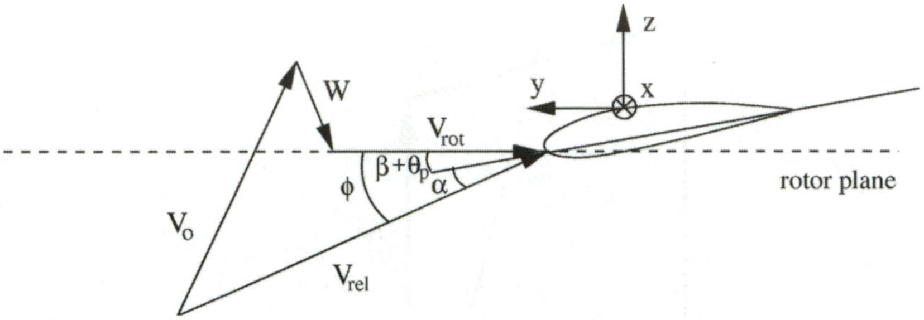

Figure 9.4 *Velocity triangle seen locally on a blade*

The angle of attack, α, can be computed if the induced velocity, $\mathbf{W}$, is known:

$$\alpha = \phi - (\beta + \theta_p), \qquad (9.10)$$

where:

$$\tan \phi = \frac{V_{rel,z}}{-V_{rel,y}} . \qquad (9.11)$$

Knowing α, the lift and drag coefficients can be looked up in a table.

The essence of the BEM method is to determine the induced velocity $\mathbf{W}$ and thus the local angles of attack.

From a global consideration the rotor acts as a disc with a discontinuous pressure drop across it. The thrust generated by this pressure drop induces a velocity normal to the rotor plane, W_n, that deflects the wake as shown in Figure 9.5.

From simple momentum theory it is known that the induced velocity in the far wake is two times the induced velocity in the rotor plane. Bramwell (1976) states that Glauert's relation between the thrust and this induced velocity for a gyrocopter in forward flight (similar to the lifting line result for an elliptically loaded circular wing) is:

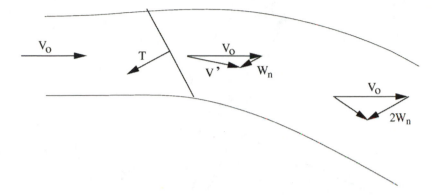

Figure 9.5 *The wake behind a rotor disc seen from above*

$$W_n = \mathbf{n} \cdot \mathbf{W} = \frac{T}{2\rho A |\mathbf{V'}|}, \tag{9.12}$$

where $|\mathbf{V'}| = |\mathbf{V_o} + \mathbf{n}(\mathbf{n} \cdot \mathbf{W})|$.

$\mathbf{n}$ is the unit vector in the direction of the thrust, which in system 3 has the coordinates $\mathbf{n} = (0,0,-1)$.

Further, equation (9.12) reduces to the classical BEM theory for zero yaw misalignment. A gyrocopter in forward flight corresponds to 90° yaw misalignment and it is therefore postulated that equation (9.12) is valid for any yaw angle, which in fact is far from obvious.

Figure 9.6 shows the local effect close to a blade section. It is assumed that only the lift contributes to the induced velocity, and that the induced velocity acts in the opposite direction to the lift.

The force from this blade at this radial position is assumed to affect the air in the area $dA = 2\pi r dr/B$, so that all B blades cover the entire annulus of the rotor disc at radius r (see Figure 9.7).

From equation (9.12) and Figure 9.6 the following can be derived for one blade:

$$W_n = W_z = \frac{-L\cos\phi \cdot dr}{2\rho \dfrac{2\pi r dr}{B} F |\mathbf{V_o} + f_g\mathbf{n}(\mathbf{n} \cdot \mathbf{W})|} = \frac{-BL\cos\phi}{4\pi\rho r F |\mathbf{V_o} + f_g\mathbf{n}(\mathbf{n} \cdot \mathbf{W})|}. \tag{9.13}$$

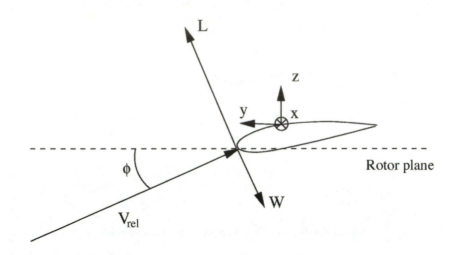

Figure 9.6 *The local effect on an airfoil*

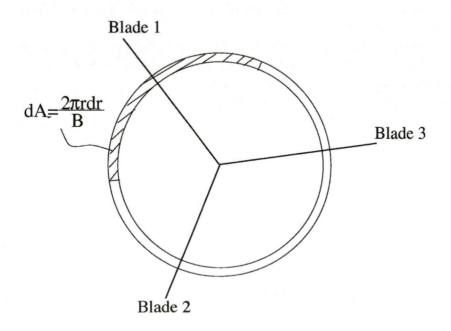

Figure 9.7 *Annular strip*

For the tangential component a similar expression is postulated:

$$\mathbf{W}_t = W_y = \frac{-BL\sin\phi}{4\pi\rho rF|\mathbf{V}_o + f_g\mathbf{n}(\mathbf{n}\cdot\mathbf{W})|} \tag{9.14}$$

F is Prandtl's tip loss factor. It is noted that the equations for the induced velocity are identical with the classical BEM method in the case of zero yaw misalignment and can be reduced to the well-known expression:

$$C_T = 4aF(1 - f_g \cdot a), \tag{9.15}$$

where by definition for an annual element of infinitesimal thickness, dr,

$$C_T = \frac{dT}{\frac{1}{2}\rho V_o^2 dA} . \tag{9.16}$$

f_g, usually referred to as the Glauert correction, is an empirical relationship between the thrust coefficient C_T and the axial induction factor a in the turbulent wake state and if equation (6.38) is used assumes the form:

$$f_g = \begin{cases} 1 & \text{for } a \le a_c \\ \dfrac{a_c}{a}(2 - \dfrac{a_c}{a}) & \text{for } a > a_c. \end{cases} \tag{9.17}$$

a_c is around 0.2. Further, it is noted that the equations must be solved iteratively since the flow angle and thus also the angle of attack depend on the induced velocity itself. But the method described here is unsteady and thus time is used as relaxation; in other words after the blades have moved in one time step an azimuthal angle of $\Delta\theta_{wing} = \omega\Delta t$ (assuming a small Δt), values from the previous time step are used on the right-hand side of equations (9.13) and (9.14) for **W** when updating new values for the induced velocity. This can be done since the induced velocity changes relatively slowly in time due to the dynamic wake model.

Dynamic Wake Model

To take into account the time delay before equations (9.13) and (9.14) are in equilibrium with the aerodynamic loads, a dynamic inflow model must be applied. In two EU-sponsored projects (Snel and Schepers, 1995; Schepers and Snel, 1995) different engineering models were tested against measurements. One of these models, proposed by S. Øye, is a filter for the induced velocities consisting of two first order differential equations:

$$W_{int} + \tau_1 \frac{dW_{int}}{dt} = W_{qs} + k \cdot \tau_1 \frac{dW_{qs}}{dt} \tag{9.18}$$

$$W + \tau_2 \frac{dW}{dt} = W_{int}. \tag{9.19}$$

W_{qs} is the quasi-static value found by equations (9.13) and (9.14), W_{int} an intermediate value and W the final filtered value to be used as the induced velocity. The two time constants are calibrated using a simple vortex method as:

$$\tau_1 = \frac{1.1}{(1-1.3a)} \cdot \frac{R}{V_o} \tag{9.20}$$

and:

$$\tau_2 = \left(0.39 - 0.26 \left(\frac{r}{R}\right)^2\right) \cdot \tau_1. \tag{9.21}$$

R is the rotor radius, $k = 0.6$ and a is the axial induction factor defined for zero yaw as $a = W_n/|V_o|$ or more generally estimated as:

$$a = \frac{|V_o|-|V'|}{|V_o|} \tag{9.22}$$

Using equation (9.20), however, a is not allowed to exceed 0.5. Equations (9.18) and (9.19) can be solved using different numerical techniques. The one suggested here is to assume that the right-hand sides are constant, which allows them to be solved analytically, yielding the following algorithm:

1 calculate W^i_{qs} using equations (9.13) and (9.14);

2 estimate right hand side of equation (9.18) using backward difference,

$$H = W^i_{qs} + k \cdot \tau_1 \frac{W^i_{qs} - W^{i-1}_{qs}}{\Delta t}$$

3 solve equation (9.18) analytically, $W^i_{int} = H + (W^{i-1}_{int} - H)\exp(\frac{-\Delta t}{\tau_1})$; and

4 solve equation (9.19) analytically, $W^i = W_{int} + (W^{i-1} - W_{int})\exp(\frac{-\Delta t}{\tau_2})$.

Applying a dynamic filter for the induced velocity is necessary in order to capture the time behaviour of the loads and power when the thrust is changed, by pitching the blades, for example. Figure 9.8 shows for a 2MW machine the computed and measured response in the rotor shaft torque for a sudden change of the pitch angle. At *t* = 2s the pitch is increased from 0 to 3.7°, decreasing the local angles of attack. First the torque drops from 260kNm to 150kNm and not until approximately 10s later do the induced velocities and thus the power settle at a new equilibrium. At *t* = 32s the pitch is changed back to 0° and a similar overshoot in the torque is observed. The decay of the spikes seen in Figure 9.8 can only be computed with a dynamic inflow model; such a model is therefore of utmost importance for a pitch regulated wind turbine.

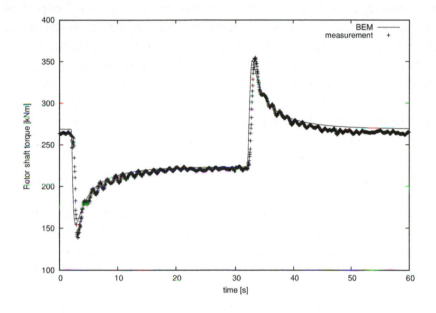

Figure 9.8 *Comparison between measured and computed time series of the rotor shaft torque for the Tjaereborg machine during a step input of the pitch for a wind speed of 8.7m/s*

Dynamic Stall

The wind seen locally on a point on the blade changes constantly due to wind shear, yaw/tilt misalignment, tower passage and atmospheric turbulence. This has a direct impact on the angle of attack, which changes dynamically during the revolution. The effect of changing the blade's angle of attack will not appear instantaneously on the loads but will take place with a time delay proportional to the chord divided by the relative velocity seen at the blade section. The response of the aerodynamic load depends on whether the boundary layer is attached or partly separated. In the case of attached flow the time delay can be estimated using the Theodorsen theory for unsteady lift and aerodynamic moment (Theodorsen, 1935). For trailing edge stall, in other words when separation starts at the trailing edge and gradually increases upstream at increasing angles of attack, so-called dynamic stall can be modelled through a separation function, f_s, as described in Øye (1991) (see later). The Beddoes–Leishman model (Leishman and Beddoes, 1989) further takes into account attached flow, leading edge separation and compressibility effects, and also corrects the drag and moment coefficients. For wind turbines, trailing edge separation is assumed to be the most important phenomenon in terms of dynamic aerofoil data, but effects in the linear region may also be important (see Hansen et al, 2004). It is shown in Øye (1991) that if a dynamic stall model is not used, one might compute flapwise vibrations, especially for stall regulated wind turbines, which are non-existent on the real machine. For stability reasons it is thus highly recommended to at least include a dynamic stall model for the lift. For trailing edge stall the degree of stall is described through f_s as:

$$C_l = f_s C_{l,inv}(\alpha) + (1 - f_s)C_{l,fs}(\alpha), \tag{9.23}$$

where $C_{l,inv}$ denotes the lift coefficient for inviscid flow without any separation and $C_{l,fs}$ is the lift coefficient for fully separated flow, for example on a flat plate with a sharp leading edge. $C_{l,inv}$ is normally an extrapolation of the static aerofoil data in the linear region; one way of estimating $C_{l,fs}$ and f_s^{st} is shown in Hansen et al (2004). f_s^{st} is the value of f_s that reproduces the static aerofoil data when applied in equation (9.23). The assumption is that f_s will always try to get back to the static value as:

$$\frac{df_s}{dt} = \frac{f_s^{st} - f_s}{\tau}, \tag{9.24}$$

which can be integrated analytically to give:

$$f_s(t + \Delta t) = f_s^{st} + (f_s(t) - f_s^{st})\exp(-\frac{\Delta t}{\tau}). \tag{9.25}$$

τ is a time constant approximately equal to $A \cdot c/V_{rel}$, where c denotes the local chord and V_{rel} is the relative velocity seen by the blade section. A is a constant that typically takes a value of about 4. Applying a dynamic stall model, the aerofoil data is always chasing the static value at a given angle of attack that is also changing in time. If, for example, the angle of attack is suddenly increased from below to above stall, the unsteady aerofoil data contains for a short time some of the inviscid/unstalled value, $C_{l,inv}$, and an overshoot relative to the static data is seen. It can thus be seen as a model of the time needed for the viscous boundary layer to develop from one state to another. Figure 9.9 shows the result using the dynamic stall model for $\alpha = 15 + 2\sin(12.57 \cdot t)$ with a time constant $\tau = 0.08$s and the initial condition $f_s(0) = f_s^{st}$. It is seen that the mean slope of the lift curve, $dC_l/d\alpha$, becomes positive for the dynamic aerofoil data in stall, which is beneficial for stability.

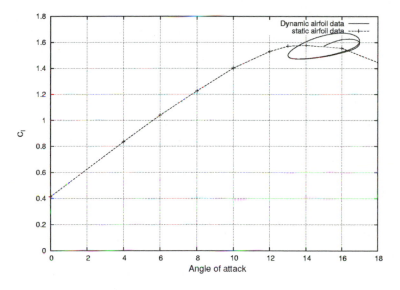

Figure 9.9 *Example of the result using a dynamic stall model*

Yaw/Tilt Model

If the rotor is yawed (and/or tilted), as shown in Figure 9.10, there will be an azimuthal variation of the induced velocity, so that the induced velocity is smaller when the blade is pointing upstream than when the same blade half a revolution later is pointing downstream.

The physical explanation of this is that a blade pointing downstream is deeper into the wake than a blade pointing upstream. This means that an upstream blade sees a higher wind and thus produces higher loads than the downstream blade, which produces a beneficial yawing moment that will try to turn the rotor more into the wind, thus enhancing yaw stability. The yaw model describes the distribution of the induced velocity. If a yaw model is

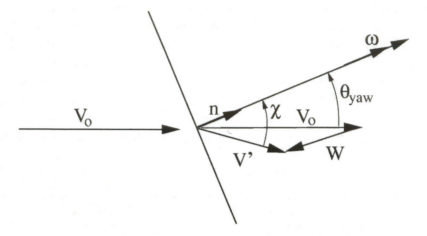

Figure 9.10 *Yawed rotor disc in wind field*

not included, the BEM method will not be able to predict the restoring yaw moment. The following yaw model proposed by Glauert is also found in Snel and Schepers (1995):

$$\mathbf{W} = \mathbf{W}_o\,(1 + \tfrac{r}{R}\tan(\tfrac{\chi}{2})\cos(\theta_{wing} -\theta_o)), \tag{9.26}$$

where the wake skew angle, χ, is defined as the angle between the wind velocity in the wake and the rotational axis of the rotor (see Figure 9.10). $\mathbf{W}_o$ is the mean induced velocity found by equations (9.13) and (9.14) and followed by (9.18) and (9.19). θ_o is the angle where the blade is deepest into the wake. The skew angle can be found as:

$$\cos\chi = \frac{\mathbf{n}\cdot\mathbf{V'}}{|\mathbf{n}||\mathbf{V'}|}, \tag{9.27}$$

where $\mathbf{n}$ is the normal vector in the direction of the rotational axis (see Figure 9.10). The skew angle is assumed to be constant with the radius and can be computed at a radial position close to $r/R = 0.7$.

The induced velocity is now known at the new azimuthal position at time $t+\Delta t$, $\theta_{wing}(t+\Delta t) = \theta_{wing}(t)+\omega\cdot\Delta t$. The angle of attack can thus be evaluated from equation (9.10) and the lift, drag and moment coefficients can be looked up from a table. The normal, p_z, and tangential, p_y, loads can be determined from:

$$p_z = L\cos\phi + D\sin\phi \tag{9.28}$$

and:

$$p_y = L\sin\phi - D\cos\phi \tag{9.29}$$

where:

$$L = \tfrac{1}{2}\rho|\mathbf{V_{rel}}|^2 cC_l \tag{9.30}$$

and:

$$D = \tfrac{1}{2}\rho|\mathbf{V_{rel}}|^2 cC_d \tag{9.31}$$

To summarize the unsteady BEM model:

- read geometry and run parameters;
- initialize the position and velocity of blades;
- discretize the blades into N elements;
- initialize the induced velocity
  - for $n = 1$ to max time step ($t = n{\cdot}\Delta t$);
  - for each blade;
  - for each element 1 to N;
- compute the relative velocity to the blade element from equation (9.9) using old values for the induced velocity;
- calculate the flow angle and thus the angle of attack from equations (9.10) and (9.11);
- determine static C_l and C_d from a table;
- determine dynamic aerofoil data using a dynamic stall model;
- calculate lift using equation (9.30);
- compute the loads p_z and p_y using equations (9.28) and (9.29);
- compute new equilibrium values for the induced velocities W_z and W_y using equations (9.13) and (9.14);
- find the unsteady induced velocities, W_z and W_y, using a dynamic wake model; and
- in the case of yaw, calculate the azimuthal variation from equation (9.26) and compute the induced velocity for each blade.

In the case of an aeroelastic computation, where the structure is not considered stiff, the loads p_z and p_y are used to determine a local blade velocity $\mathbf{V_b}$. This blade velocity must be taken into account when computing the relative velocity and equation (9.9) should be extended to:

$$\begin{pmatrix} V_{rel,y} \\ V_{rel,z} \end{pmatrix} = \begin{pmatrix} V_y \\ V_z \end{pmatrix} + \begin{pmatrix} -\omega x \cos\theta_{cone} \\ 0 \end{pmatrix} + \begin{pmatrix} W_y \\ W_z \end{pmatrix} - \begin{pmatrix} V_{b,y} \\ V_{b,z} \end{pmatrix} \qquad (9.32)$$

Deterministic Model for Wind

The time averaged atmospheric boundary layer shown in Figure 9.11 can be modelled as:

$$V_o(x) = V_o(H) \left(\frac{x}{H}\right)^v, \qquad (9.33)$$

where H is the hub height, x the distance from the surface and v a parameter giving the amount of shear. v is in the range between 0.1 and 0.25.

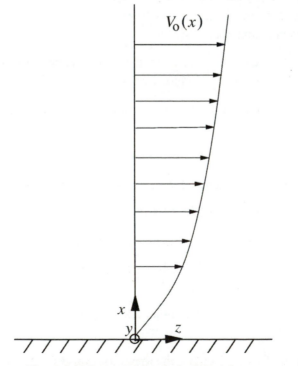

Figure 9.11 *Deterministic wind velocity shear*

The wind is also influenced by the presence of the tower. A simple model for the influence of the tower is to assume potential flow (see Figure 9.12).

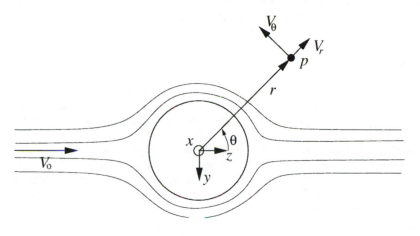

Figure 9.12 *The effect of the tower*

Coordinate system 1 is oriented so that the undisturbed wind velocity, V_o, is aligned with the z-axis. In a polar coordinate system as shown in Figure 9.12, the radial and tangential velocity components around the tower, assuming potential flow, can be computed as:

$$V_r = V_o(1-(\tfrac{a}{r})^2)\cos\theta, \tag{9.34}$$

and:

$$V_\theta = -V_o(1+(\tfrac{a}{r})^2)\sin\theta, \tag{9.35}$$

where a is the radius of the tower. Transforming the velocity to the cartesian coordinate system 1 yields:

$$\begin{aligned} V_z &= V_r \cos\theta - V_\theta \sin\theta \\ V_y &= -V_r \sin\theta - V_\theta \cos\theta \end{aligned} \tag{9.36}$$

The following relation between r, θ and y, z can be used:

$$\begin{aligned} \cos\theta &= \tfrac{z}{r} \\ \sin\theta &= -\tfrac{y}{r}. \\ r &= \sqrt{z^2 + y^2} \end{aligned} \tag{9.37}$$

It is noted that assuming potential flow is a bad approximation for a downwind machine, where each blade passes the tower wake once every revolution. Further, the turbulent part of the real atmospheric wind should be added for a realistic time simulation of a wind turbine.

References

Bramwell, A. R. S. (1976) *Helicopter Dynamics*, Edward Arnold Ltd, London

Hansen, M. H., Gaunaa, M. and Madsen, H. A. (2004) 'A Beddoes–Leishman type dynamic stall model in state-space and indicial formulations', Risoe-R-1354(EN), Roskilde, Denmark

Leishman, J. G. and Beddoes, T. S. (1989) 'A semi-empirical model for dynamic stall', *Journal of the American Helicopter Society*, vol 34, no 3, pp3–17

Øye, S. (1991) 'Dynamic stall, simulated as a time lag of separation', in K. F. McAnulty (ed) *Proceedings of the 4th IEA Symposium on the Aerodynamics of Wind Turbines*, ETSU-N-118, Harwell Laboratory, Harwell, UK

Schepers, J. G. and Snel, H. (1995) *Dynamic Inflow: Yawed Conditions and Partial Span Pitch Control*, ECN-C- -95-056, Petten, The Netherlands

Snel, H. and Schepers, J. G. (1995) *Joint Investigation of Dynamic Inflow Effects and Implementation of an Engineering Method,* ECN-C- -94-107, Petten, The Netherlands

Theodorsen, T. (1935) 'General theory of aerodynamic instability and the mechanism of flutter', NACA report no 496, National Advisory Committee for Aeronautics, pp413–433

10

Introduction to Loads and Structures

Having described in detail how to calculate the aerodynamic loads on a wind turbine, the following text concerns the structural issues that need to be addressed to ensure that the construction will not break down during its design lifetime of typically 20 years. Normally a breakdown is caused by an inadequate control system, extreme wind conditions, fatigue cracks or a defective safety system. A very dangerous breakdown may occur if power to the generator is lost. In this case there is no braking torque on the rotor, which, in the absence of a safety system such as mechanical or aerodynamic emergency brakes, is free to accelerate. Because the aerodynamic forces increase with the square of the rotor speed, the blades will bend more and more in the downwind direction and might end up hitting the tower or flying off due to centrifugal forces. It has been estimated (Sørensen, 1983) that torn-off blades from an overspeeding wind turbine could land up to about 300 m from the tower. Fortunately, violent failures are extremely rare and no humans have, to the author's knowledge, ever been reported to have suffered injuries from debris flying off a wind turbine. Safety standards such as IEC 61400 (2004) exist to ensure that wind turbines operate safely. The standards define load cases, such as extreme gusts, which a wind turbine must be able to survive. Lightning is also known to have caused disintegration of blades.

Fatigue is a very important issue in a wind turbine construction, which is built to run for a minimum of 20 years and thus performs in the order of 10^9 revolutions. To estimate the loads on a wind turbine throughout its lifetime, the loads and hence the stresses in the material must either be computed using an aeroelastic code in a realistic wind field including turbulence or be measured directly on an already built turbine. Once the dynamic stresses are known, it is possible to calculate the fatigue damage using standard methods such as the Palmgren-Miner rule.

Description of Main Loads

In order that the following text can be better understood, a short description of the main loads on a horizontal-axis wind turbine is given. To extract energy from the wind it is necessary to slow down the wind speed using a force pointing in the upwind direction. This force is called the thrust and is caused by a jump in the pressure over the rotor, induced by the flow past the individual rotor blades. The total load has not only a component normal to the flow but also a tangential component in the rotational direction of the blades. The tangential load component delivers the shaft torque that turns the rotor. To characterize these loads it is common to state the flapwise and the edgewise bending moments at a position close to the root of the blades together with the tilt rotor moment and yaw rotor moment in the shaft between the rotor and the first bearing. The flapwise bending moment M_{flap} (see Figure 10.1) stems from the normal forces (thrust), which tend to deflect the blades out of the rotor plane in the downwind direction:

$$M_{flap}(r_{pos}) = \int_{r_{pos}}^{R} rp_N(r)dr, \tag{10.1}$$

where p_N is the normal force per length, r the local radius, R the total radius of the rotor and dr an incremental part of the blade. The edgewise bending moment is the bending moment in the rotor plane from the tangential force distribution. The edgewise bending moment is sometimes referred to as the lead-lag moment.

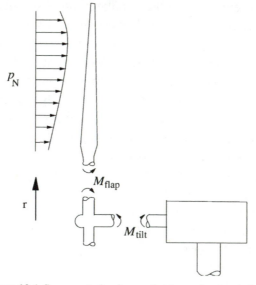

Figure 10.1 *Some main loads on a horizontal-axis wind turbine*

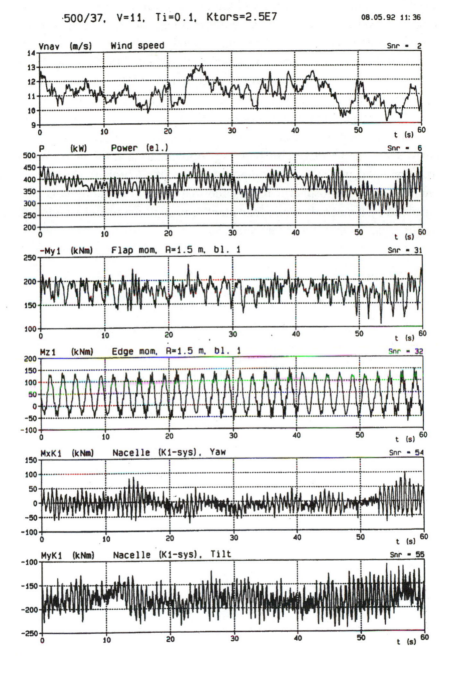

Figure 10.2 *Example of computed loads using FLEX4 for a mean wind speed of 11m/s*

The tilt rotor moment in the main shaft, as shown in Figure 10.1, tries to tilt the nacelle over the tower. The yaw rotor moment tries to turn the nacelle on the tower. Sometimes the two bending moments at the root of the tower are also stated. Figure 10.2 shows an example of computed loads using the aero-elastic code FLEX4 at a mean wind speed of 11m/s. The wind speed at hub height and the power output are seen on the first graphs. The last four graphs show the corresponding time histories of the flapwise and edgewise bending moments and the tilt rotor and yaw rotor moments. The effect of gravity is clearly seen on the edgewise bending moment as a dominant sinusoidal variation, upon which is superimposed some small high frequency signal stemming from atmospheric turbulence. The flapwise bending is mostly influenced by the aerodynamic loads that vary with the turbulent wind field, and this signal is therefore more stochastic.

References

IEC 61400-1 (2004) 'Wind turbines. Part 1: Design requirements', CD, edition 3, second revision, IEC TC88-MT1

Sørensen, J. N. (1983) 'Beregning af banekurver og kastelængder for afrevne vindmøllevinger' (in Danish), AFM83-06, Department of Fluid Mechanics (AFM), Technical University of Denmark (DTU)

11

Beam Theory for the Wind Turbine Blade

This section describes how a blade, whose outer contour is designed from aerodynamical considerations, is built to be sufficiently strong and stiff. In the past materials like wood, steel, aluminium, glass-fibre-reinforced plastics (GRPs) and carbon fibre reinforced plastics (CFRPs) have been used. The choice depends on many parameters such as strength, weight, stiffness, price and, very important for wind turbines, fatigue properties. The majority of wind turbine blades today are built using GRPs, and therefore a short description of a manufacturing process using this material is given. A negative mould for the upper part (suction side) and lower part (pressure side) of the blade is made. A thin film of so-called gelcoat is first laid in the moulds. The gelcoat gives a smooth white finish to the blades and therefore it is not necessary to paint the blades afterwards. Then a number of glass fibre mats are laid in. On each mat a layer of epoxy or polyester is rolled on to bind the mats into a hard matrix of fibres. The number of mats gives the thickness of the shell; typically a thin shell is made around the leading and trailing edges and a thick shell is made in the middle of the aerofoil. A section of such a blade is shown in Figure 11.1.

To make the blade stronger and stiffer, so-called webs are glued on between the two shells before they are glued together. To make the trailing edge stiffer, foam panels can also be glued on before assembling the upper and lower parts. Because such a construction consists of different layers, it is often called a sandwich construction; a sketch is given in Figure 11.2 which can readily be compared to the real section in Figure 11.1. It is seen that the thick layer of mats and epoxy in the middle of the skin and the webs form a box-like structure. For structural analysis, the box-like structure, which is the most important structural part of the blade, acts like a main beam on which a thin skin is glued defining the geometry of the blade. Fixing a thin skin on a main beam (so the skin is not carrying loads, but merely gives an outer aerodynamic shape) is an alternative way that is sometimes used to construct a blade.

Figure 11.1 *Section of an actual blade*

A blade can thus be modelled as a beam, and when the stiffnesses *EI* and *GI$_v$* at different spanwise stations are computed, simple beam theory can be applied to compute the stresses and deflections of that blade. *E* is the modulus of elasticity, *G* is the modulus of elasticity for shear and *I* denotes different moments of inertia. In the next section a more elaborate explanation is given for the moments of inertia and it is shown how the stiffnesses can be computed for a wind turbine blade such as the one shown in Figure 11.1.

The simple beam theory described here is found in almost any basic book on mechanics of materials (for example Gere and Timoshenko, 1972). Further, it is outlined how to compute the important structural parameters shown in Figure 11.3. Values of these parameters are necessary to compute the deflection of a blade for a given load or as input to a dynamic simulation using an aeroelastic code.

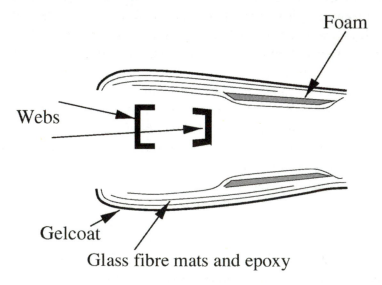

Figure 11.2 *Schematic drawing of a section of a blade*

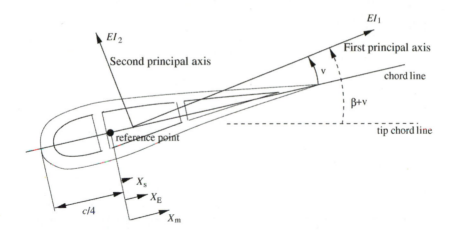

Figure 11.3 *Section of a blade showing the main structural parameters*

EI_1 – bending stiffness about first principal axis;
EI_2 – bending stiffness about second principal axis;
GI_v – torsional stiffness;
X_E – the distance of the point of elasticity from the reference point;
X_m – the distance of the centre of mass from the reference point;
X_s – the distance of the shear centre from the reference point;
β – the twist of the aerofoil section measured relative to the tip chord
      line;
v – angle between chord line and first principal axis;
$\beta+v$ – angle between tip chord line and first principal axis;

The point of elasticity is defined as the point where a normal force (out of the plane) will not give rise to a bending of the beam. The shear centre is the point where an in-plane force will not rotate the aerofoil. If the beam is bent about one of the principal axes, the beam will only bend about this axis. As will be seen later, it is convenient to use the principal axes when calculating the blade deflection.

Before continuing, some necessary definitions must be introduced. The following quantities are defined in terms of the reference coordinate system (X_R, Y_R) in Figure 11.4:

- Longitudinal stiffness: $[EA] = \int_A E dA$.

- Moment of stiffness about the axis X_R: $[ES_{X_R}] = \int_A EY_R dA$.

- Moment of stiffness about the axis Y_R: $[ES_{Y_R}] = \int_A EX_R dA$.

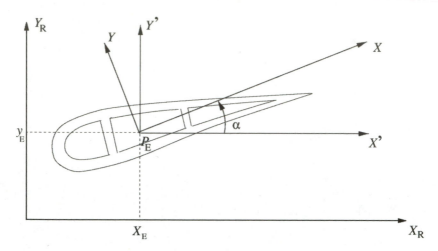

Figure 11.4 *Section of a blade*

- Moment of stiffness inertia about the axis X_R: $[EI_{X_R}] = \int_A EY_R^2 dA$.

- Moment of stiffness inertia about the axis Y_R: $[EI_{Y_R}] = \int_A EX_R^2 dA$.

- Moment of centrifugal stiffness: $[ED_{XY_R}] = \int_A EX_R Y_R dA$.

From these definitions, the point of elasticity $P_E = (X_E, Y_E)$ can be computed in the reference system (X_R, Y_R) as:

$$X_E = \frac{[ES_{Y_R}]}{[EA]} \tag{11.1}$$

and:

$$Y_E = \frac{[ES_{X_R}]}{[EA]} \tag{11.2}$$

For E and ρ constant the point (X_E, Y_E) equals the centre of mass for the section, where ρ denotes the density of the material used. Now the moments of stiffness inertia and the moment of centrifugal stiffness are moved to the coordinate system (X', Y'), which is parallel to the reference system (X_R, Y_R) and has its origin in the point of elasticity, using the formulaes:

$$[EI_X] = \int_A E(Y')^2 dA = [EI_{X_R}] - Y_E^2 [EA] \tag{11.3}$$

$$[EI_{Y'}] = \int_A E(X')^2 dA = [EI_{Y_R}] - X_E^2[EA] \tag{11.4}$$

$$[ED_{X'Y'}] = \int_A EX'Y'dA = [ED_{XY_R}] - X_E Y_E[EA]. \tag{11.5}$$

It is now possible to compute the angle α between X' and the first principal axis and the bending stiffness about the principal axes. The second principal axis is perpendicular to the first principal axis:

$$\alpha = \frac{1}{2}\tan^{-1}\left(\frac{2[ED_{X'Y'}]}{[EI_{Y'}] - [EI_{X'}]}\right) \tag{11.6}$$

$$[EI_1] = [EI_{X'}] - [ED_{X'Y'}]\tan\alpha \tag{11.7}$$

$$[EI_2] = [EI_{Y'}] + [ED_{X'Y'}]\tan\alpha \tag{11.8}$$

The stress $\sigma(x,y)$ in the cross-section from the bending moments about the two principal axes M_X and M_Y and the normal force N is found from:

$$\sigma(x,y) = E(x,y)\varepsilon(x,y), \tag{11.9}$$

where the strain ε is computed from:

$$\varepsilon(x,y) = \frac{M_1}{[EI_1]} y - \frac{M_2}{[EI_2]} x + \frac{N}{[EA]}. \tag{11.10}$$

σ, ε and N are positive for tension and negative for compression. The bending moments M_1 and M_2 and the normal force N are computed from the loading of the blade, as is shown later.

The main structural data are now determined. Since a wind turbine blade is very stiff in torsion, the torsional deflection is normally not considered. A complete description of how to compute the shear centre and the torsional rigidity is, however, given in Øye (1978). An example of results from Øye (1988), for the 30m blade used at the 2MW Tjæreborg wind turbine, is listed in Table 11.1.

Table 11.1 shows that the position of the first principal axis, described by the angle $\beta+v$, varies with the radius r. It is also seen that the position of the first principal axis is almost identical with the chord line since the angle v is small for most of the blade.

Table 11.1 *Main structural parameters for the Tjaereborg blade*

r [m]	EA [GN]	EI_1 [MNm²]	EI_2 [MNm²]	GI_v [MNm²]	m [kg/m]	X_E [mm]	X_m [mm]	X_s [mm]	v [°]	$\beta+v$ [°]
1.8	36.0	12,000	12,000	7500	1700	0	0	0	0	0
3.0	6.14	1630	1725	362	330	2	2	0	5.4	14.4
4.5	5.82	1080	1940	328	389	54	159	11	0.94	9.44
6.0	5.10	623	1490	207	347	59	165	13	1.30	9.30
9.0	4.06	255	905	92.8	283	63	170	18	1.09	8.09
12.0	3.33	129	557	47.7	235	58	158	15	0.86	6.86
15.0	2.76	64.8	349	24.7	196	51	137	15	0.86	5.86
18.0	2.33	32.4	221	12.9	166	45	121	16	0.91	4.91
21.0	1.83	15.2	131	6.23	172	41	110	17	0.83	3.83
24.0	1.21	6.04	65.7	2.57	90.3	40	102	16	0.63	2.63
27.0	0.63	1.82	28.1	0.84	52.6	47	108	14	0.16	1.16
30.0	0.21	0.32	9.5	0.18	24.2	82	136	10	–0.52	–0.52

Deflections and Bending Moments

A wind turbine blade such as that in Figure 11.5 can be treated as a technical beam as sketched in Figure 11.6.

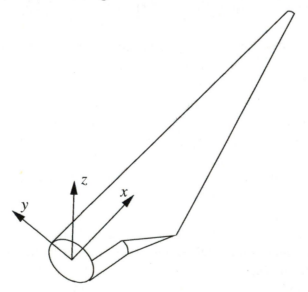

Note that the coordinate system used for the blade is different from the one shown in Figure 11.4.

Figure 11.5 *Wind turbine blade*

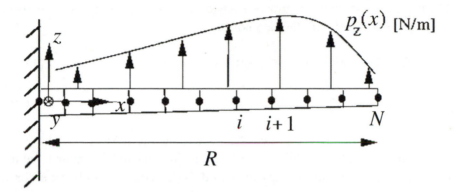

Figure 11.6 *Technical beam*

If the external loadings, p_y and p_z, are known along the blade, the shear forces, T_z and T_y, and bending moments, M_y and M_z, can be found as:

$$\frac{dT_z}{dx} = -p_z(x) + m(x)\ddot{u}_z(x) \tag{11.11}$$

$$\frac{dT_y}{dx} = -p_y(x) + m(x)\ddot{u}_y(x) \tag{11.12}$$

$$\frac{dM_y}{dx} = T_z \tag{11.13}$$

$$\frac{dM_z}{dx} = -T_y. \tag{11.14}$$

Equations (11.11–11.14) can be derived using Newton's second law on an infinitesimal piece of the beam as shown in Figure 11.7. $\ddot{u}$ is the acceleration and if the blade is in equilibrium, the last term (inertia term) on the right hand side of equations (11.11) and (11.12) is zero.

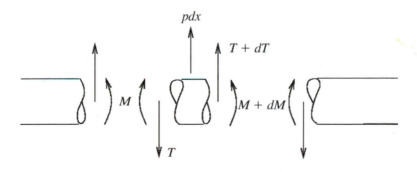

Figure 11.7 *Infinitesimal piece of the beam*

The bending moments can now be transformed to the principal axes as (here the y-axis is aligned with the tip chord):

$$M_1 = M_y \cos(\beta + v) - M_z \sin(\beta + v) \qquad (11.15)$$

$$M_2 = M_y \sin(\beta + v) + M_z \cos(\beta + v), \qquad (11.16)$$

where β+v is the angle between the y-axis and the first principal axis as shown in Figure 11.8. All other things being equal, it can be assumed that the first principal axis lies along the chord line, which is only true for a symmetric aerofoil. Note that β is negative for a normally twisted blade, but (β+v) is assumed positive in equations (11.15), (11.16), (11.19) and (11.20).

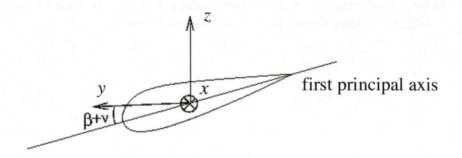

Figure 11.8 *Orientation of principal axes*

The curvatures about the principal axes are from simple beam theory:

$$\kappa_1 = \frac{M_1}{EI_1} \qquad (11.17)$$

$$\kappa_2 = \frac{M_2}{EI_2}. \qquad (11.18)$$

These curvatures are then transformed back to the y-axis and z-axis as:

$$\kappa_z = -\kappa_1 \sin(\beta + v) + \kappa_2 \cos(\beta + v) \qquad (11.19)$$

$$\kappa_y = \kappa_1 \cos(\beta + v) + \kappa_2 \sin(\beta + v) \qquad (11.20)$$

The angular deformations and thus the deflections are now calculated from:

$$\frac{d\theta_y}{dx} = \kappa_y \tag{11.21}$$

$$\frac{d\theta_z}{dx} = \kappa_z \tag{11.22}$$

$$\frac{du_z}{dx} = -\theta_y \tag{11.23}$$

$$\frac{du_y}{dx} = \theta_z. \tag{11.24}$$

If the loads are given in discrete points as shown in Figure 11.9 and the loads are assumed to vary linearly between two points i and $i+1$, then equations (11.11) to (11.14) and equations (11.21) to (11.24) can be integrated to give the following numerical algorithm to calculate bending moments and deflections.

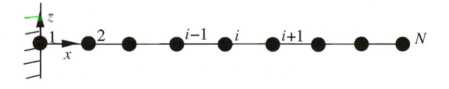

Figure 11.9 *Discretized cantilever beam*

Numerical Algorithm for Determining the Bending Moments and Deflection

Boundary conditions: $T_y^N = 0$ and $T_z^N = 0$

For $i = N$ to 2:

$$T_y^{i-1} = T_y^i + \frac{1}{2}(p_y^{i-1} + p_y^i)(x^i - x^{i-1}) \tag{11.25}$$

$$T_z^{i-1} = T_z^i + \frac{1}{2}(p_z^{i-1} + p_z^i)(x^i - x^{i-1}) \tag{11.26}$$

Boundary conditions: $M_y^N = 0$ and $M_z^N = 0$

For $i = N$ to 2:

$$M_y^{i-1} = M_y^i - T_z^i(x^i - x^{i-1}) - (\tfrac{1}{6}p_z^{i-1} + \tfrac{1}{3}p_z^i)(x^i - x^{i-1})^2 \tag{11.27}$$

$$M_z^{i-1} = M_z^i + T_y^i(x^i - x^{i-1}) + (\tfrac{1}{6}p_y^{i-1} + \tfrac{1}{3}p_y^i)(x^i - x^{i-1})^2. \tag{11.28}$$

For all points calculate M_1 and M_2 using equations (11.15) and (11.16). Then κ_y and κ_z are computed using equations (11.17–11.20).

Boundary conditions: $\theta^1_y = 0$ and $\theta^1_z = 0$

For $i = 1,N–1$:

$$\theta_y^{i+1} = \theta_y^i + \tfrac{1}{2}(\kappa_y^{i+1} + \kappa_y^i)(x^{i+1} - x^i) \tag{11.29}$$

$$\theta_z^{i+1} = \theta_z^i + \tfrac{1}{2}(\kappa_z^{i+1} + \kappa_z^i)(x^{i+1} - x^i). \tag{11.30}$$

Boundary conditions: $u^1_y = 0$ and $u^1_z = 0$

For $i = 1,N–1$:

$$u_y^{i+1} = u_y^i + \theta_z^i(x^{i+1} - x^i) + (\tfrac{1}{6}\kappa_z^{i+1} + \tfrac{1}{3}\kappa_z^i)(x^{i+1} - x^i)^2 \tag{11.31}$$

$$u_z^{i+1} = u_z^i - \theta_y^i(x^{i+1} - x^i) - (\tfrac{1}{6}\kappa_y^{i+1} + \tfrac{1}{3}\kappa_y^i)(x^{i+1} - x^i)^2. \tag{11.32}$$

The boundary conditions are for a cantilever beam and the inertia terms in equations (11.11) and (11.12) have been neglected for simplicity, but must be added for an unsteady computation.

A Method to Estimate the First Flapwise, First Edgewise and Second Flapwise Eigenmodes

The equations (11.25) to (11.32) can also be used to estimate the first few eigenmodes. An eigenmode is a free vibration without the presence of external loads; in other words equations (11.11) and (11.12) become:

$$\frac{dT_z}{dx} = m(x)\ddot{u}_z(x) \tag{11.33}$$

$$\frac{dT_y}{dx} = m(x)\ddot{u}_y(x). \tag{11.34}$$

Since for an eigenmode the deflection is of the type $u = A\sin(\omega t)$, the acceleration is proportional to the deflection as:

$$\ddot{u} = -\omega^2 u, \tag{11.35}$$

where ω is the associated eigenfrequency. Using (11.35), equations (11.33) and (11.34) become:

$$\frac{dT_z}{dx} = -m(x)\omega^2 u_z(x) \tag{11.36}$$

$$\frac{dT_y}{dx} = -m(x)\omega^2 u_y(x). \tag{11.37}$$

Comparing equations (11.11–11.12) with (11.36–11.37), it is seen that an eigenmode can be found using the static beam equations applying the external loads as:

$$p_z = m(x)\omega^2 u_z(x) \tag{11.38}$$

$$p_y = m(x)\omega^2 u_y(x). \tag{11.39}$$

Since the deflections in equations (11.38) and (11.39) are not known, the equations must be solved iteratively, which will converge to the mode with the lowest eigenfrequency, known as the first flapwise mode. First, an initial deflection can be found using, for example, a constant loading for both the z and y directions. With this deflection the eigenfrequency is estimated at the tip as:

$$\omega^2 = \frac{p_z^N}{u_z^N m^N} \tag{11.40}$$

and a new loading is computed in all discrete points as:

$$p_z^i = \omega^2 m^i \frac{u_z^i}{\sqrt{(u_z^N)^2 + (u_y^N)^2}}. \tag{11.41}$$

$$p_y^i = \omega^2 m^i \frac{u_y^i}{\sqrt{(u_z^N)^2 + (u_y^N)^2}}. \tag{11.42}$$

Note that the loading is normalized with the tip deflection to ensure that the tip deflection becomes 1 in the next iteration. A new deflection is now computed using the loading from equations (11.41) and (11.42). The procedure is repeated a few times until the eigenfrequency ω becomes constant. Now the deflection shape, u_z^{1f} and u_y^{1f}, of the first flapwise eigenmode as sketched in Figure 11.10 is known.

To find the first edgewise deflection the procedure can be used again. However, some modification is needed to make sure the method does *not* converge towards the first flapwise mode. Every time a new deflection, $u_y(x)$ and $u_z(x)$, is computed, it is necessary to subtract the part that contains the first flapwise mode :

$$u_z^{1e} = u_z - const_1 \cdot u_z^{1f} \tag{11.43}$$

$$u_y^{1e} = u_y - const_1 \cdot u_y^{1f}. \tag{11.44}$$

To determine the constant, the following orthogonality constraint must be satisfied:

$$\int_0^R u_z^{1f} mu_z^{1e} dx + \int_0^R u_y^{1f} mu_y^{1e} dx = 0 \tag{11.45}$$

Combining equations (11.43) and (11.44) with equation (11.45) yields an expression for the constant:

$$const_1 = \frac{\int_0^R u_z^{1f} mu_z dx + \int_0^R u_y^{1f} mu_y dx}{\int_0^R u_z^{1f} mu_z^{1f} dx + \int_0^R u_y^{1f} mu_y^{1f} dx}. \tag{11.46}$$

Thus to force the method to converge to the first edgewise eigenmode it is necessary every time a new deflection has been computed to remove the part from the first flapwise eigenmode using equations (11.43), (11.44) and (11.46). After a few iterations the deflection shape of the first edgewise eigenmode, u_z^{1e} and u_y^{1e}, as sketched in Figure 11.11, is known.

Finally, the second flapwise eigenmode can be found similarly. Now it is also necessary to substract not only the part that contains the first flapwise eigenmode but also the part from the first edgewise eigenmode using the following orthogonality constraint:

$$\int_0^R u_z^{1e} mu_z^{2f} dx + \int_0^R u_y^{1e} mu_y^{2f} dx = 0. \tag{11.47}$$

Every time a new deflection has been computed, equations (11.43), (11.44) and (11.46) are used to remove the part from the first flapwise eigenmode. Next equations (11.48–11.50) are applied to remove the part from the first edgewise eigenmode fulfilling constraint (11.47):

$$u_z^{2f} = u_z - const_2 \cdot u_z^{1e} \tag{11.48}$$

$$u_y^{2f} = u_y - const_2 \cdot u_y^{1e} \tag{11.49}$$

$$const_2 = \frac{\int_0^R u_z^{1e} m u_z dx + \int_0^R u_y^{1e} m u_y dx}{\int_0^R u_z^{1e} m u_z^{1e} dx + \int_0^R u_y^{1e} m u_y^{1e} dx} \tag{11.50}$$

After a few iterations the deflection shape of the second flapwise eigenmode, u_z^{2f} and u_y^{2f}, as sketched in Figure 11.12, is known.

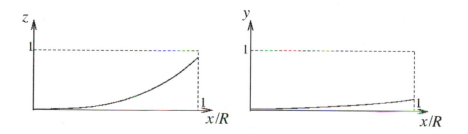

Figure 11.10 *First flapwise eigenmode (1f)*

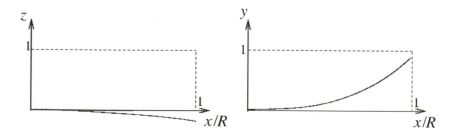

Figure 11.11 *First edgewise eigenmode (1e)*

The eigenmode estimation can be checked by comparing with the analytical solutions for a cantilever beam having constant properties along the beam (see, for example, Craig, 1981):

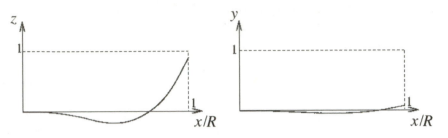

Figure 11.12 *Second flapwise eigenmode (2f)*

$$\omega_1 = \frac{3.516}{L^2} \left(\frac{EI}{m} \right)^{\frac{1}{2}} \tag{11.51}$$

$$\omega_2 = \frac{22.03}{L^2} \left(\frac{EI}{m} \right)^{\frac{1}{2}} \tag{11.52}$$

ω_1 and ω_2 are the first and second eigenfrequencies respectively and L is the length of the beam. Applying this algorithm on a beam with constant stiffness $EI = 1\text{Nm}^2$, a mass distribution $m = 1\text{kg/m}$, a length $L = 1\text{m}$ and using 11 points yields $\omega_1 = 3.513\text{rad/s}$, which is close to the analytical solution of 3.516 rad/s. Using 11 points, the second eigenfrequency becomes 22.273rad/s and 22.044rad/s for 51 points, which should be compared to the analytical value of 22.03rad/s. The solution for 51 points is shown in Figure 11.13.

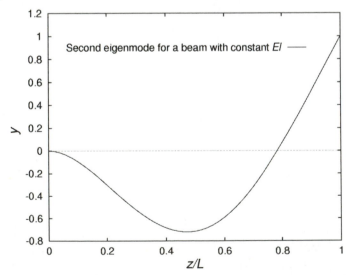

Figure 11.13 *Second eigenmode using 50 elements*

In an accurate computation the effect of the centrifugal force on a deflected blade must be included. In Figure 11.14, the total force F_x from the centrifugal acceleration at $r = x$ is shown.

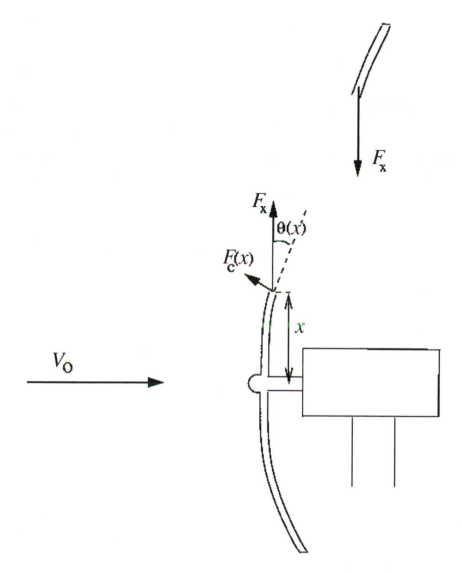

Figure 11.14 *Total centrifugal force* F$_x$ *at a spanwise position* x

F_x can be computed by integrating the incremental contribution dF_x from x to the tip of the blade R.

$$F_x = \int_x^R dF_x = \int_x^R x'\omega^2 m dx', \tag{11.53}$$

where ω is the angular velocity of the rotor, m the mass per length, dx' an incremental part of the blade and the x' radial distance to the incremental part. The projection of F_x normal to the rotor blade is (see Figure 11.14):

$$F_c = F_x \sin\theta \approx F_x\theta. \tag{11.54}$$

This force corresponds to a loading p_c (force per length):

$$p_c = \frac{dF_c}{dx} = F_x \frac{d\theta}{dx} + \theta \frac{dF_x}{dx} \tag{11.55}$$

F_x decreases for increasing x, thus $dF_x/dx = -x\omega^2 m$ and equation (11.55) becomes:

$$p_c = F_x \frac{d\theta}{dx} - mx\omega^2\theta \tag{11.56}$$

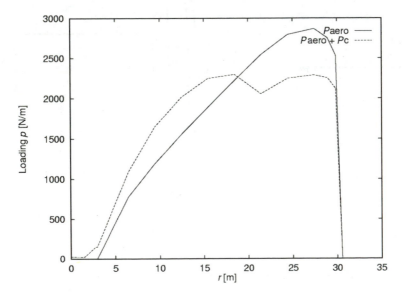

p_{aero} is the aerodynamic load computed by BEM method and $p_{aero} + p_c$ is the aerodynamic plus the centrifugal load.

Figure 11.15 *The computed load about the first principal axis for the 2MW Tjæreborg machine at* $V_o = 10m/s$ *with and without the centrifugal loading*

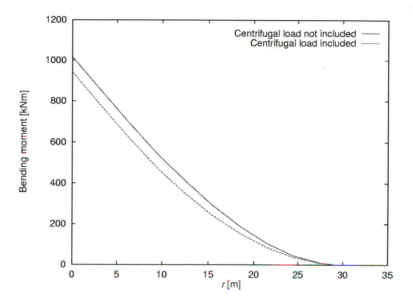

Figure 11.16 *The computed bending moment about the first principal axis for the 2MW Tjæreborg machine at* $V_o = 10m/s$ *with and without the centrifugal loading*

The extra loading from equation (11.56) should be added to the load calculated from the aerodynamics and the net result is a redistribution of the loading, which results in a reduction in the flapwise bending moment. Since p_c is a function of the angular deformation θ and the curvature $d\theta/dx$, which again is a function of the loads, it is necessary to iterate a few times in order to have the deflection including the effect of the centrifugal acceleration. In other words, first the deflections and the angular deformations are computed by using only the aerodynamic loads. Then the extra loading is computed using equations (11.53) and (11.56). A new loading is found by adding this extra loading to the original aerodynamic loading and a new set of deflections and angular deformations are computed. Hereafter the centrifugal loading is calculated again using equation (11.56) and the procedure is repeated a few times until a stationary solution is found. Figure 11.15 shows the effect of including the centrifugal loading on the 2MW Tjæreborg machine at a wind speed of 10m/s. It is seen that the centrifugal loading reduces the loads at the tip but increases the loads at the root. The corresponding reduction in the bending moments is seen in Figure 11.16.

References

Craig, R. R. Jr (1981) *Structural Dynamics*, John Wiley & Sons, Hoboken, NJ

Gere, J. M. and Timoshenko, S. P. (1972) *Mechanics of Materials*, van Nostrand Reinhold Company, New York

Øye, S. (1978) 'Beregning af tværsnitskonstanter for bjælker med inhomogent tværsnit' (in Danish), AFM notat VK-45-781201, Department of Fluid Mechanics, Technical University of Denmark (DTU)

Øye, S. (1988) 'Projekt K 30m Glasfibervinge, Teknisk beskrivelse' (in Danish), AFM 88-12, Department of Fluid Mechanics, Technical University of Denmark (DTU)

12

Dynamic Structural Model
of a Wind Turbine

The main purpose of a structural model of a wind turbine is to be able to determine the temporal variation of the material loads in the various components in order to estimate the fatigue damage. Further, a dynamic system is used when analysing the stability of the wind turbine design, including perhaps the control system. To calculate the deflections and velocities of the various components in the wind turbine in the time domain, a structural model including the inertia terms is needed. Then the dynamic structural response of the entire construction can be calculated subject to the time dependent load found using an aerodynamic model, such as the BEM method. For offshore wind turbines, wave loads and perhaps ice loads on the bottom of the tower must also be estimated. One way of setting up the structural model, based on the principle of virtual work, is presented here in detail. However, more formal 'finite element' methods have also been used in different aeroelastic codes. The velocity of the vibrating wind turbine construction must be subtracted when calculating the relative velocity seen locally by the blade as shown in equation (9.32). The loads therefore depend on the deflections and velocities of the structure, which again depend on the loads. The structural and aerodynamic models are therefore highly coupled and must be solved together in what is known as an aeroelastic problem.

Principle of Virtual Work and Use
of Modal Shape Functions

The principle of virtual work is a method to set up the correct mass matrix, $\mathbf{M}$, stiffness matrix, $\mathbf{K}$, and damping matrix, $\mathbf{C}$, for a discretized mechanical system as:

$$\mathbf{M\ddot{x}} + \mathbf{C\dot{x}} + \mathbf{Kx} = \mathbf{F}_g, \tag{12.1}$$

where $\mathbf{F_g}$ denotes the generalized force vector associated with the external loads, $\mathbf{p}$. Equation (12.1) is of course nothing but Newton's second law, assuming linear stiffness and damping, and the method of virtual work is nothing but a method that helps in setting up the correct mass, stiffness and damping matrices for a multi-body system, which is especially well suited for a chain system. Knowing the loads and appropriate conditions for the velocities and the deformations, equation (12.1) can be solved for the accelerations, from which the velocities and deformations can be estimated for the next time step. The number of elements in $\mathbf{x}$ is called the number of degrees of freedom, DOF, and the higher this number the more computational time is needed in each time step to solve the matrix system. Use of modal shape functions is a tool to reduce the number of degrees of freedom and thus reduce the size of the matrices to make the computations faster per time step. A deflection shape in this method is described as a linear combination of a few but physically realistic basis functions, which are often the deflection shapes corresponding to the eigenmodes with the lowest eigenfrequencies. For a wind turbine such an approach is suited to describe the deflection of the rotor blades and the assumption is that the combination of the power spectral density of the loads and the damping of the system do not excite eigenmodes associated with higher frequencies. In the commercially available and widely used aeroelastic simulation tool FLEX (see, for example, Øye, 1996), only the first 3 or 4 (2 flapwise and 1 or 2 edgewise) eigenmodes are used for the blades and results are in good agreement with measurements, indicating the validity of the assumption. First one has to decide on the DOFs necessary to describe a realistic deformation of a wind turbine. For instance, in FLEX4 17–20 DOFs are used for a three- bladed wind turbine: 3–4 DOFs per blade as described above; the deformation of the shaft is described using 4 DOFs (1 for torsion, 2 hinges just before the first bearing with associated angular stiffness to describe bending and 1 for pure rotation); 1 DOF to describe the tilt stiffness of the nacelle; and finally 3 DOFs for the tower (1 for torsion, 1 mode in the normal direction of the rotor and 1 in the lateral direction).

The values in the vector $\mathbf{x}$ describing the deformation of the construction, x_i, are known as the general coordinates. To each generalized coordinate is associated a deflection shape, $\mathbf{u}_i$, that describes the deformation of the construction when only x_i is different from zero and typically has a unit value. The element i in the generalized force corresponding to a small displacement in DOF number i, dx_i, is calculated such that the work done by the generalized force equals the work done on the construction by the distributed external loads on the associated deflection shape:

$$F_{g,i}dx_i = \int_s \mathbf{p \cdot u}_i dS, \tag{12.2}$$

where S denotes the entire system. Please note that the generalized force can be a moment and that the displacement can be angular. All loads must be included, in other words also gravity and inertial loads such as Coriolis, centrifugal and gyroscopic loads. The non-linear centrifugal stiffening can be modelled as equivalent loads calculated from the local centrifugal force and the actual deflection shape as shown in the previous chapter. The elements in the mass matrix, $m_{i,j}$, can be evaluated as the generalized force from the inertia loads from an unit acceleration of DOF j for a unit displacement of DOF i. The elements in the stiffness matrix, $k_{i,j}$, correspond to the generalized force from an external force field which keeps the system in equilibrium for a unit displacement in DOF j and which then is displaced $x_i = 1$. The elements in the damping matrix can be found similarly. For a chain system the method of virtual work as described here normally gives a full mass matrix and diagonal matrices for the stiffness and damping. For one blade rigidly clamped at the root (cantilever beam), it is relatively easy to estimate the lowest eigenmodes (first flapwise $\mathbf{u}^{1f}$, first edgewise $\mathbf{u}^{1e}$ and second flapwise $\mathbf{u}^{2f}$), using, for example, the iterative procedure from the previous chapter. This description of the principle of virtual work might seem very abstract, but will hopefully become clearer after using it on two examples, one of two discrete masses connected by springs and dampers and another for an isolated blade (see later).

If the structural system comprises a system of continuous mass distributions, such as a system of beams, equation (12.1) is the result of discretizing the system, since in reality such a system has an infinite number of DOFs. The elements in the mass, stiffness and damping matrices depend on the system and, in the case of a continuous system, also of the discretization. If the right hand side of equation (12.1) is 0 the system is said to perform its natural motion.

Provided that the deflections, $\mathbf{x}$, and velocities, $\mathbf{\dot{x}}$, are known, equation (12.1) can alternatively be written as:

$$\mathbf{M\ddot{x} = F_g - C\dot{x} - Kx = f(\dot{x}, x, \mathit{t})}, \tag{12.3}$$

where the function $\mathbf{f}$ in general is non-linear. Non-linarity can come, for example, from non-linear loads $\mathbf{p}$ or from aerodynamic damping (see later). A non-linear system can be treated as a linearized eigenvalue approach or as a full non-linear time domain approach. Only the latter method is treated in this text.

Knowing the right hand side of equation (12.3) at time $t^n = n\Delta t$, the acceleration at time t^n is found solving the linear system of equations:

$$\ddot{\mathbf{x}}^n = \mathbf{M}^{-1}\mathbf{f}(\dot{\mathbf{x}}^n, \mathbf{x}^n, t^n).$$

(12.4)

Knowing the accelerations, $\ddot{\mathbf{x}}^n$, the velocities, $\dot{\mathbf{x}}^n$, and positions, $\mathbf{x}^n$, at time t^n, a Runge-Kutta-Nyström scheme can be used to estimate the velocities, $\dot{\mathbf{x}}^{n+1}$, and positions, $\mathbf{x}^{n+1}$, at t^{n+1}. New loads, $\mathbf{p}^{n+1}(\dot{\mathbf{x}}^{n+1}, \mathbf{x}^{n+1}, t^{n+1})$, can be calculated using, for example, an unsteady BEM method and thus equation (12.4) can be updated and a new time step can be performed. This can be continued until a sufficient time period has been simulated.

Box 12.1 *The Runge-Kutta-Nyström integration scheme of* $\ddot{\mathbf{x}} = \mathbf{g}(t, \dot{\mathbf{x}}, \mathbf{x})$

$$\mathbf{A} = \frac{\Delta t}{2}\ddot{\mathbf{x}}^n$$

$$\mathbf{b} = \frac{\Delta t}{2}(\dot{\mathbf{x}}^n + \tfrac{1}{2}\mathbf{A})$$

$$\mathbf{B} = \frac{\Delta t}{2}\mathbf{g}(t^{n+\frac{1}{2}}, \mathbf{x}^n + \mathbf{b}, \dot{\mathbf{x}}^n + \mathbf{A})$$

$$\mathbf{C} = \frac{\Delta t}{2}\mathbf{g}(t^{n+\frac{1}{2}}, \mathbf{x}^n + \mathbf{b}, \dot{\mathbf{x}}^n + \mathbf{B})$$

$$\mathbf{d} = \Delta t(\dot{\mathbf{x}}^n + \mathbf{C})$$

$$\mathbf{D} = \frac{\Delta t}{2}\mathbf{g}(t^{n+1}, \mathbf{x}^n + \mathbf{d}, \dot{\mathbf{x}}^n + 2\mathbf{C})$$

and the final update:

$$t^{n+1} = t^n + \Delta t$$

$$\mathbf{x}^{n+1} = \mathbf{x}^n + \Delta t(\dot{\mathbf{x}}^n + \tfrac{1}{3}(\mathbf{A} + \mathbf{B} + \mathbf{C}))$$

$$\dot{\mathbf{x}}^{n+1} = \dot{\mathbf{x}}^n + \tfrac{1}{3}(\mathbf{A} + 2\mathbf{B} + 2\mathbf{C} + \mathbf{D})$$

$$\ddot{\mathbf{x}}^{n+1} = \mathbf{g}(t^{n+1}, \mathbf{x}^{n+1}, \dot{\mathbf{x}}^{n+1})$$

SDOF

The simplest dynamic system is called SDOF (single degree of freedom) and is comprised of only one concentrated mass. Imagine, for example, a mass which is hung up in a spring as shown in Figure 12.1.

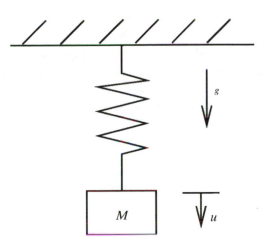

Figure 12.1 *SDOF system with no damping*

Statically the spring will be stretched until the spring force equals the weight Mg. The dynamic equation is:

$$M\ddot{u} + ku = 0, \tag{12.5}$$

where k is the spring constant and u the displacement from the equilibrium position. The well known analytical solution to this system is:

$$u = u_o \cos(\omega_N t) + \frac{\dot{u}_o}{\omega_N} \sin(\omega_N t). \tag{12.6}$$

u_o is the pertubation at $t=0$, $\dot{u}_o$ is the velocity at $t = 0$ and $\omega_N = \sqrt{k/M}$ is the eigenfrequency of the system. The system performs a simple harmonic vibration and the time for one cycle is $T = 2\pi/\omega_N$. This simple SDOF system can be used, for example, to check the implementation, accuracy and stability of the Runge-Kutta-Nyström method.

Aerodynamic Damping

The aerodynamic forces from the flow past a structure can cause damping that may be negative and thus amplify vibrations. Below is a description of the aerodynamic damping using a simple SDOF system, but now including aerodynamic forces, as sketched in Figure 12.2. The system comprises an aerofoil in a wind tunnel, which is mounted with a geometrical angle of attack α_g on a spring that allows the aerofoil to move up and down. Imagine now that the blade is moving downwards with a velocity $\dot{x}$. The aerofoil then feels an extra velocity, equal to $\dot{x}$ but in the opposite direction, giving a component from below that, when added to the velocity in the wind tunnel V_o, gives the relative velocity (see Figure 12.2). Provided that the aerofoil data, $C_l(\alpha)$, are known, the flow angle ϕ, the angle of attack and the force in the x-direction can be estimated as:

$$\tan \phi = \frac{\dot{x}}{V_o} \tag{12.7}$$

$$\alpha = \alpha_g + \phi \tag{12.8}$$

$$F_x = \tfrac{1}{2}\rho V_{rel}^2 \, AC_l(\alpha)\cos\phi \tag{12.9}$$

When the aerofoil moves downwards, the angle of attack increases, and when it moves upwards, the angle of attack is decreased according to equations (12.7) and (12.8). This changes the lift coefficient:

$$C_l = C_{l,o} + \frac{\partial C_l}{\partial \alpha}\, \Delta\alpha, \tag{12.10}$$

where $C_{l,o}$ is the previous value of the lift coefficient and C_l the new lift coefficient at the higher angle of attack. If the aerofoil is moving downwards and the slope $\partial C_l/\partial\alpha$ is positive, the lift coefficient and thus the aerodynamic force is increased and, since this increased force is in the opposite direction to the motion, the vibration is damped. The same argument is valid when the blade is moving upwards. If the slope is negative, however, as in stalled conditions, the aerodynamic damping is negative. More formally, the work done by the aerodynamic forces on the section during one cycle can be calculated as:

$$W = -\oint \mathbf{F}\cdot\mathbf{dx}. \tag{12.11}$$

$\mathbf{F}$ is the aerodynamic force on the aerofoil and $\mathbf{x}$ the displacement. If the work is positive the section is positively damped, and if the work is negative the

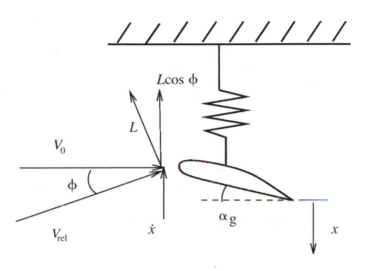

Figure 12.2 *SDOF with lift*

section is negatively damped. To have the correct slopes for the lift and drag coefficients is therefore essential for correctly predicting the stability. For these types of vibrations, dynamic stall models as described in Chapter 9 are thus very important. In some cases the dynamic stall model effectively increases the slope $\partial C_l/\partial\alpha$, and thus the stability. In other words computations may over-predict the oscillations if static aerofoil data are used.

Examples of Using the Principle of Virtual Work

Two examples are given here of how to use the principle of virtual work to create the mass, stiffness and damping matrices in equation (12.1). The first example is a 2-DOF system comprising two masses connected by springs and dampers as shown in Figure 12.3. The second example is a wind turbine blade using modal shapes in order to reduce the number of DOFs.

2-DOF system

The mass and stiffness matrices together with the generalized force vector for the 2-DOF system shown in Figure 12.3 will be set up using the principle of virtual work and generalized coordinates.

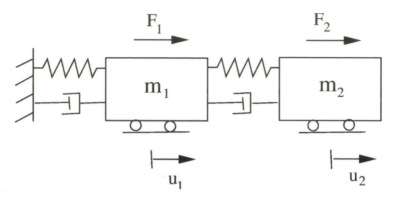

Figure 12.3 *Example of a 2-DOF system*

The methodology is as follows: first, the generalized coordinates, (x_1, x_2), are defined as the relative displacements between the two masses:

$$x_1 = u_1$$
$$x_2 = u_2 - u_1$$

(12.12)

where u_1 and u_2 are the displacements of mass 1 and 2, respectively. The generalized force vector, $\mathbf{F}_g$, is found by the principle of virtual work as the work done by the external forces, F_1 and F_2, for a displacement of one of the generalized coordinates keeping the other(s) zero. The first component is found for $x_1 = 1$ and $x_2 = 0$ as $F_{g,1} = F_1 + F_2$ since in this case both masses moves a unit length. The second component is found putting $x_1 = 0$ and $x_2 = 1$ as $F_{g,2} = F_2$ since in this case only mass 2 moves a unit length. The mass matrix is found by specifying a unit acceleration of one of the generalized coordinates keeping the other(s) zero and replacing the external forces by inertia forces, in other words mass times acceleration. For $\ddot{x}_1 = 1$ and $\ddot{x}_2 = 0$ both masses 1 and 2 accelerate with a unit acceleration, in other words the corresponding inertial forces becomes $F_1 = m_1$ and $F_2 = m_2$ and the generalized force based on these become $F_{g,1} = m_1 + m_2$ and $F_{g,2} = m_2$. This gives the first column in the mass matrix as shown below:

$$\begin{bmatrix} m_{11} & m_{12} \\ m_{21} & m_{22} \end{bmatrix} \begin{bmatrix} 1 \\ 0 \end{bmatrix} = \begin{bmatrix} m_{11} \\ m_{21} \end{bmatrix} = \begin{bmatrix} m_1 + m_2 \\ m_2 \end{bmatrix}.$$

(12.13)

The second column is found by specifying $\ddot{x}_1 = 0$ and $\ddot{x}_2 = 1$; in other words only mass 2 accelerates and the inertia forces become $F_1 = 0$ and $F_2 = m_2$. The generalized force based on these inertia forces becomes $F_{g,1} = m_2$ and $F_{g,2} = m_2$, yielding:

$$\begin{bmatrix} m_{11} & m_{12} \\ m_{21} & m_{22} \end{bmatrix} \begin{bmatrix} 0 \\ 1 \end{bmatrix} = \begin{bmatrix} m_{12} \\ m_{22} \end{bmatrix} = \begin{bmatrix} m_2 \\ m_2 \end{bmatrix}. \qquad (12.14)$$

The mass matrix therefore becomes:

$$\mathbf{M} = \begin{bmatrix} m_1 + m_2 & m_2 \\ m_2 & m_2 \end{bmatrix} \qquad (12.15)$$

The first column in the stiffness matrix can be found using the necessary generalized force to obtain a unit static displacement of the first generalized coordinate, i.e. $x_1 = 1$ and $x_2 = 0$. The necessary external forces for this unit displacement are $F_1 = k_1$ and $F_2 = 0$, where k_1 is the spring constant for spring 1. The corresponding generalized force is $F_{g,1} = F_1 + F_2 = k_1$ and $F_{g,2} = F_2 = 0$ and the first column in the stiffness matrix becomes:

$$\begin{bmatrix} k_{11} & k_{12} \\ k_{21} & k_{22} \end{bmatrix} \begin{bmatrix} 1 \\ 0 \end{bmatrix} = \begin{bmatrix} k_{11} \\ k_{21} \end{bmatrix} = \begin{bmatrix} k_1 \\ 0 \end{bmatrix}. \qquad (12.16)$$

The second column is found as the necessary generalized force to obtain a unit static displacement of the second generalized coordinate, i.e. $x_1 = 0$ and $x_2 = 1$. The necessary external forces for this unit displacement are $F_1 = -k_2$ and $F_2 = k_2$, where k_2 is the spring constant for spring 2. The corresponding generalized force is $F_{g,1} = F_1 + F_2 = 0$ and $F_{g,2} = F_2 = k_2$, yielding:

$$\begin{bmatrix} k_{11} & k_{12} \\ k_{21} & k_{22} \end{bmatrix} \begin{bmatrix} 0 \\ 1 \end{bmatrix} = \begin{bmatrix} k_{12} \\ k_{22} \end{bmatrix} = \begin{bmatrix} 0 \\ k_2 \end{bmatrix}. \qquad (12.17)$$

The stiffness matrix therefore becomes:

$$\mathbf{K} = \begin{bmatrix} k_1 & 0 \\ 0 & k_2 \end{bmatrix}. \qquad (12.18)$$

The damping matrix can be found similarly. The first column is the generalized force necessary to obtain a unit velocity of the first generalized coordinate, i.e. $\dot{x}_1 = 1$ and $\dot{x}_2 = 0$. The result is similar to the stiffness matrix and the damping matrix becomes:

$$\mathbf{C} = \begin{bmatrix} d_1 & 0 \\ 0 & d_2 \end{bmatrix}. \tag{12.19}$$

where d_1 and d_2 are the coefficients of viscous damping for the two dampers. In a chain formulation like the above example the stiffness and damping matrices become diagonal matrices, whereas the mass matrix becomes full. The full system is thus determined as:

$$\begin{bmatrix} m_1 + m_2 & m_2 \\ & m_2 & m_2 \end{bmatrix} \begin{bmatrix} \ddot{x}_1 \\ \ddot{x}_2 \end{bmatrix} + \begin{bmatrix} d_1 & 0 \\ 0 & d_2 \end{bmatrix} \begin{bmatrix} \dot{x}_1 \\ \dot{x}_2 \end{bmatrix} + \begin{bmatrix} k_1 & 0 \\ 0 & k_2 \end{bmatrix} \begin{bmatrix} x_1 \\ x_2 \end{bmatrix} = \begin{bmatrix} F_1 + F_2 \\ F_2 \end{bmatrix} \tag{12.20}$$

To verify the described method Newton's second law is applied on each mass:

$$m_1 \ddot{u}_1 = -k_1 u_1 - d_1 \dot{u}_1 + k_2(u_2 - u_1) + d_2(\dot{u}_2 - \dot{u}_1) + F_1 \tag{12.21}$$

$$m_2 \ddot{u}_2 = -k_2(u_2 - u_1) - d_2(\dot{u}_2 - \dot{u}_1) + F_2. \tag{12.22}$$

Replacing u_1 by x_1 and u_2 by $x_1 + x_2$ and replacing equation (12.21) with equation (12.21) + equation (12.22) makes equations (12.21) and (12.22) identical to equation system (12.20). It is straightforward to add more masses.

Dynamic system for blade

The methodology of generalized coordinates is now applied to a wind turbine blade as shown in Figure 11.5. Assume that we know the normalized eigenmodes, with a maximum tip deflection of 1m. The first eigenmodes can then be calculated as shown in the previous chapter and sketched in Figures 11.10–11.12.

It is assumed that a deformation of a blade can be described as a linear combination of these three modes as:

$$u_z(x) = x_1 \cdot u_z^{1y}(x) + x_2 \cdot u_z^{1e}(x) + x_3 \cdot u_z^{2f}(x) \tag{12.23}$$

and:

$$u_y(x) = x_1 \cdot u_y^{1y}(x) + x_2 \cdot u_y^{1e}(x) + x_3 \cdot u_y^{2f}(x) \tag{12.24}$$

The deflection shape can thus be described by three parameters only, which are denoted by the generalized coordinates x_1, x_2 and x_3. Since the modes are constant, the velocities and accelerations along the blade are:

$$\dot{u}_z(x) = \dot{x}_1 \cdot u_z^{1f}(x) + \dot{x}_2 \cdot u_z^{1e}(x) + \dot{x}_3 \cdot u_z^{2f}(x) \tag{12.25}$$

$$\dot{u}_y(x) = \dot{x}_1 \cdot u_y^{1f}(x) + \dot{x}_2 \cdot u_y^{1e}(x) + \dot{x}_3 \cdot u_y^{2f}(x) \tag{12.26}$$

and:

$$\ddot{u}_z(x) = \ddot{x}_1 \cdot u_z^{1f}(x) + \ddot{x}_2 \cdot u_z^{1e}(x) + \ddot{x}_3 \cdot u_z^{2f}(x) \tag{12.27}$$

$$\ddot{u}_y(x) = \ddot{x}_1 \cdot u_y^{1f}(x) + \ddot{x}_2 \cdot u_y^{1e}(x) + \ddot{x}_3 \cdot u_y^{2f}(x) \tag{12.28}$$

As indicated by equation (12.2), the generalized force for each mode is the work done on this mode by the external loads, $p_z(x)$ and $p_y(x)$, without contribution from the other modes, i.e.:

$$F_{g,1} = \int p_z(x)u_z^{1f}(x)dx + \int p_y(x)u_y^{1f}(x)dx \tag{12.29}$$

$$F_{g,2} = \int p_z(x)u_z^{1e}(x)dx + \int p_y(x)u_y^{1e}(x)dx \tag{12.30}$$

and:

$$F_{g,3} = \int p_z(x)u_z^{2f}(x)dx + \int p_y(x)u_y^{2f}(x)dx \tag{12.31}$$

The first column of the mass matrix is found by evaluating the generalized force from external forces corresponding to the inertia forces for a unit acceleration of the first degree of freedom and the others set to 0, i.e. $(\ddot{x}_1, \ddot{x}_2, \ddot{x}_3) = (1, 0, 0)$. Using equations (12.27) and (12.28), the inertia loads become $(p_y, p_z) = (m\ddot{u}_y, m\ddot{u}_z) = (mu_y^{1f}, mu_z^{1f})$ and thus:

$$\begin{bmatrix} m_{11} \\ m_{21} \\ m_{31} \end{bmatrix} = \begin{bmatrix} \int u_z^{1f}(x)m(x)u_z^{1f}(x)dx + \int u_y^{1f}(x)m(x)u_y^{1f}(x)dx \\ \int u_z^{1f}(x)m(x)u_z^{1e}(x)dx + \int u_y^{1f}(x)m(x)u_y^{1e}(x)dx \\ \int u_z^{1f}(x)m(x)u_z^{2f}(x)dx + \int u_y^{1f}(x)m(x)u_y^{2f}(x)dx \end{bmatrix} = \begin{bmatrix} GM_1 \\ 0 \\ 0 \end{bmatrix}. \tag{12.32}$$

The first element is sometimes denoted by the first generalized mass GM_1; the two other integrals are 0 due to the orthogonality constraints between eigenmodes.

The second column of the mass matrix is found by evaluating the generalized force from external forces corresponding to the inertia forces for a unit acceleration of the second degree of freedom and the others set to 0, i.e. $(\ddot{x}_1, \ddot{x}_2, \ddot{x}_3) = (0, 1, 0)$. Using equations (12.27) and (12.28) the inertia loads become $(p_y, p_z) = (m\ddot{u}_y, m\ddot{u}_z) = (mu_y^{1e}, mu_z^{1e})$ and thus:

$$
\begin{bmatrix} m_{12} \\ m_{22} \\ m_{32} \end{bmatrix} = \begin{bmatrix} \int u_z^{1e}(x)m(x)u_z^{y}(x)dx + \int u_y^{1e}(x)m(x)u_y^{y}(x)dx \\ \int u_z^{1e}(x)m(x)u_z^{1e}(x)dx + \int u_y^{1e}(x)m(x)u_y^{1e}(x)dx \\ \int u_z^{1e}(x)m(x)u_z^{2y}(x)dx + \int u_y^{1e}(x)m(x)u_y^{2y}(x)dx \end{bmatrix} = \begin{bmatrix} 0 \\ GM_2 \\ 0 \end{bmatrix}. \quad (12.33)
$$

The second element is sometimes denoted the second generalized mass; the two other integrals are 0 due to the orthogonality constraints between eigenmodes.

The third column of the mass matrix is found by evaluating the generalized force from external forces corresponding to the inertia forces for a unit acceleration of the third degree of freedom and the others set to 0, i.e. $(\ddot{x}_1, \ddot{x}_2, \ddot{x}_3) = (0, 0, 1)$. Using equations (12.27) and (12.28), the inertia loads become $(p_y, p_z) = (m\ddot{u}_y, m\ddot{u}_z) = (mu_y^{2y}, mu_z^{2y})$ and thus:

$$
\begin{bmatrix} m_{13} \\ m_{23} \\ m_{33} \end{bmatrix} = \begin{bmatrix} \int u_z^{2y}(x)m(x)u_z^{y}(x)dx + \int u_y^{2y}(x)m(x)u_y^{y}(x)dx \\ \int u_z^{2y}(x)m(x)u_z^{1e}(x)dx + \int u_y^{2y}(x)m(x)u_y^{1e}(x)dx \\ \int u_z^{2y}(x)m(x)u_z^{2y}(x)dx + \int u_y^{2y}(x)m(x)u_y^{2y}(x)dx \end{bmatrix} = \begin{bmatrix} 0 \\ 0 \\ GM_3 \end{bmatrix}. \quad (12.34)
$$

The third element is sometimes denoted the third generalized mass; the two other integrals are 0 due to the orthogonality constraints between eigenmodes.

The first column in the stiffness matrix can be found using the generalized force necessary to obtain a unit static displacement of the first generalized coordinate, i.e. $(x_1, x_2, x_3) = (1, 0, 0)$. In this case the deflection, according to equations (12.23) and (12.24), is identical to u_y^{y}, u_z^{y}. The loads yielding this deflection, according to equations (11.38) and (11.39), are $(p_y, p_z) = (m\omega_1^2 u_y^{y}, m\omega_1^2 u_z^{y},)$, where ω_1 is the eigenfrequency associated with the first flapwise eigenmode; applying this in the equations for the generalized force (12.29–12.31) yields:

$$
\begin{bmatrix} k_{11} \\ k_{21} \\ k_{31} \end{bmatrix} = \begin{bmatrix} \int \omega_1^2 u_z^{y}mu_z^{y}dx + \int \omega_1^2 u_y^{y}mu_y^{y}dx \\ \int \omega_1^2 u_z^{y}mu_z^{1e}dx + \int \omega_1^2 u_y^{y}mu_y^{1e}dx \\ \int \omega_1^2 u_z^{y}mu_z^{2y}dx + \int \omega_1^2 u_y^{y}mu_y^{2y}dx \end{bmatrix} = \begin{bmatrix} \omega_1^2 GM_1 \\ 0 \\ 0 \end{bmatrix}. \quad (12.35)
$$

The two last integrals are 0 due to the orthogonality constraint of the eigenmodes. Similarly it can be shown that:

$$
\begin{bmatrix} k_{12} \\ k_{22} \\ k_{32} \end{bmatrix} = \begin{bmatrix} 0 \\ \omega_2^2 GM_2 \\ 0 \end{bmatrix} \tag{12.36}
$$

and:

$$
\begin{bmatrix} k_{13} \\ k_{23} \\ k_{33} \end{bmatrix} = \begin{bmatrix} 0 \\ 0 \\ \omega_3^2 GM_3 \end{bmatrix}. \tag{12.37}
$$

ω_2 and ω_3 are the eigenfrequencies associated with the first edgewise and second flapwise eigenmodes respectively. The generalized masses GM_1, GM_2 and GM_3 are defined in equations (12.32–12.34).

Neglecting the structural damping, the equation for one blade becomes:

$$
\begin{bmatrix} GM_1 & 0 & 0 \\ 0 & GM_2 & 0 \\ 0 & 0 & GM_3 \end{bmatrix} \begin{bmatrix} \ddot{x}_1 \\ \ddot{x}_2 \\ \ddot{x}_3 \end{bmatrix} + \begin{bmatrix} \omega_1^2 GM_1 & 0 & 0 \\ 0 & \omega_2^2 GM_2 & 0 \\ 0 & 0 & \omega_3^2 GM_3 \end{bmatrix} \begin{bmatrix} x_1 \\ x_2 \\ x_3 \end{bmatrix} = \begin{bmatrix} F_{g,1} \\ F_{g,2} \\ F_{g,3} \end{bmatrix}. \tag{12.38}
$$

Structural damping terms can be modelled as:

$$
\mathbf{C} = \begin{bmatrix} \omega_1 GM_1 \dfrac{\delta_1}{\pi} & 0 & 0 \\ 0 & \omega_2 GM_2 \dfrac{\delta_2}{\pi} & 0 \\ 0 & 0 & \omega_3 GM_3 \dfrac{\delta_3}{\pi} \end{bmatrix}. \tag{12.39}
$$

where δ_i is logarithmic decrement associated with mode i. The system for the beam in equation (12.38) comprises three uncoupled differential equations. This is a result of the orthogonality constraints on the eigenmodes used; it is not a general result of the principle of virtual work. For instance, when the method is used on a whole wind turbine construction the equation of motion becomes fully coupled.

FEM Models

Alternatively, a more formal finite element method (FEM) can be used to set up the dynamic structural model in the form of equation (12.1). In Ahlström (2005) and Schepers (2002) a long list of various aeroelastic codes are compiled, and many of the recently developed codes have used the FEM approach. However, the number of DOFs in a FEM discretization will be considerably larger than, for example, that from combining the principle of virtual work with mode shapes as shown in the text above for a single blade. Therefore the computational time required to calculate a time history of a certain length is much larger when a FEM approach is taken. In this book there will be no attempt to use the FEM approach for modelling a wind turbine, but more details and further reference can be found in Hansen et al (2006), for example.

References

Ahlström, A. (2005) 'Aeroelastic simulation of wind turbine dynamics', Doctoral thesis in structural mechanics, Royal Institute of Technology, KTH, Stockholm, Sweden

Hansen, M. O. L., Sørensen, J. N., Voutsinas, S., Sørensen, N. and Madsen, H. Aa. (2006) 'State of the art in wind turbine aerodynamics and aeroelasticity', *Progress in Aerospace Sciences*, vol 42, no 4, pp285–330

Øye, S. (1996) 'FLEX4 simulation of wind turbine dynamics' in *Proceedings of 28th IEA Meeting of Experts Concerning State of the Art of Aeroelastic Codes for Wind Turbine Calculations* (available through International Energy Agency)

Schepers, J. G. (2002) 'Verification of European wind turbine design codes, VEWTDC; final report', technical report ECN-C-01-055, Energy Research Centre of The Netherlands

13

Sources of Loads on
a Wind Turbine

The three most important sources of the loading of a wind turbine are

1 gravitational loading;

2 inertial loading; and

3 aerodynamic loading.

Gravitational Loading:

The Earth's gravitational field causes a sinusoidal gravitational loading on each blade, as indicated in Figure 13.1.

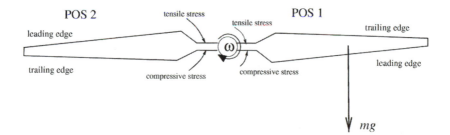

Figure 13.1 *The loading caused by the Earth's gravitational field*

When the blade is in position 1 in Figure 13.1 (down-rotating) the blade root at the trailing edge side is exposed to tensile stress and the leading edge side of the blade root is exposed to compressive stress. In position 2 (up-rotating) the trailing edge side of the blade root is exposed to compressive stress and the leading edge side of the blade root is exposed to tensile stress. Thus gravity is responsible for a sinusoidal loading of the blades with a frequency corresponding to the rotation of the rotor often denoted by 1P. This loading is easily recognized in Figure 10.2 in the time series of the edgewise bending moment. Note that a wind turbine is designed to operate for 20 years, which means that a machine operating at 25 rpm will be exposed to $20 \times 365 \times 24 \times 60 \times 25 = 2.6 \times 10^8$ stress cycles from gravity. Since a wind turbine blade might weigh several tons and be more than 30m long, the stresses from the gravity loading are very important in the fatigue analysis.

Inertial Loading

Inertial loading occurs when, for example, the turbine is accelerated or decelerated. An example is the braking of the rotor, where a braking torque T is applied at the rotor shaft. A small section of the blade will feel a force dF in the direction of the rotation as indicated in Figure 13.2.

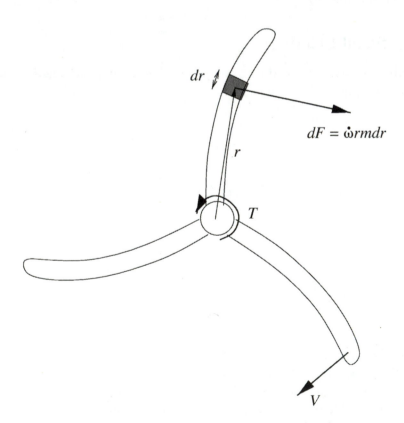

Figure 13.2 *Loading caused by braking the rotor*

The size of *dF* is found from:

$$dF = \dot{\omega}rm\,dr, \tag{13.1}$$

where *m* is the mass per length of the blade, *r* the radius from the rotational axis to the section and *dr* the size of the small section; $\dot{\omega} = d\omega/dt$ can be found from:

$$I\frac{d\omega}{dt} = T, \tag{13.2}$$

where *I* is the moment of inertia of the rotor. The *mü* terms in equations (11.11) and (11.12) are also inertia loads stemming from local accelerations. Another inertial loading stems from the centrifugal force on the blades. In order to reduce the flapwise bending moment, the rotor can be coned backwards with a cone angle of θ_{cone} as shown in Figure 13.3.

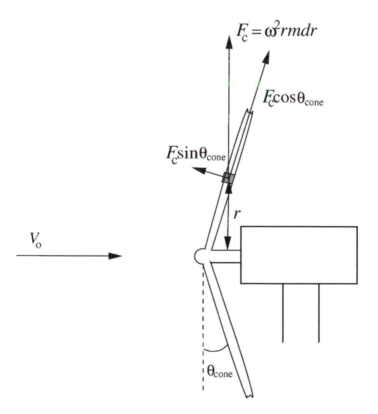

Figure 13.3 *Effect of coning the rotor*

The centrifugal force acting on the incremental part of the blade at a radius r from the rotational axis as shown in Figure 13.3 is $F_c = \omega^2 rm dr$, where M is the mass of the incremental part and ω the angular velocity of the rotor. Due to the coning the centrifugal force has a component in the spanwise direction of the blade, $F_c\cos\theta_{cone}$, and a component normal to the blade, $F_c\sin\theta_{cone}$, as shown in Figure 13.3. The normal component gives a flapwise bending moment in the opposite direction to the bending moment caused by the thrust and thus reduces the total flapwise bending moment.

Aerodynamic Loading

The aerodynamic loading is caused by the flow past the structure, in other words the blades and the tower. The wind field seen by the rotor varies in space and time due to atmospheric turbulence as sketched in Figure 13.4.

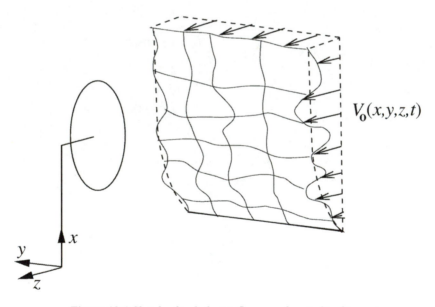

$V_0(x,y,z,t)$

Figure 13.4 *Sketch of turbulent inflow seen by wind turbine rotor*

As also seen in Figure 13.4, the wind field is characterized by shear; in other words the mean wind speed increases with the height above the ground. For neutral stability this shear may be estimated as:

$$\frac{V_{10min}(x)}{V_{10min}(h)} = \frac{\ln(x/z_o)}{\ln(h/z_o)}. \tag{13.3}$$

$V_{10min}(x)$ is the time averaged value for a period of 10 minutes at a height x above the ground. $V_{10min}(h)$ is the time averaged value at a fixed height h and z_o is the so-called roughness length. Alternatively the wind shear can be given by an exponent as in equation (9.33). The roughness length depends on the surface characteristics and varies from 10^{-4}m over water to approximately 1m in cities. Values of z_o can be found in Troen and Petersen (1989) and are summarized in Table 13.1.

Table 13.1 *Roughness length table*

z_o [m]	Terrain surface characteristics
1.0	City
0.8	Forest
0.2	Many trees and bushes
0.1	Farmland with closed appearance
0.05	Farmland with open appearance
0.03	Farmland with very few buildings, trees, etc.
5×10^{-3}	Bare soil
1×10^{-3}	Snow surface
3×10^{-4}	Sand surface (smooth)
1×10^{-4}	Water areas

Source: Troen and Petersen (1989).

Wind shear gives a sinusoidal variation of the wind speed seen by a blade with a frequency corresponding to the rotation of the rotor 1P. The turbulent fluctuations superimposed on the mean wind speed also produce a time variation in the wind speed and thus in the angle of attack. In order to simulate the behaviour of a wind turbine using an aeroelastic code, it is therefore necessary first to generate a realistic wind field as shown in Chapter 14. The tower also contributes to variation in the inflow; for an upwind rotor this can be calculated using equations (9.34–9.36).

A wind turbine might operate in yaw if, for example, the direction of the wind is not measured correctly or in the event of a malfunctioning yaw system. In this case, the reduced wind speed u at the rotor plane has a component normal to the rotor, $u\cos\theta_{yaw}$, and tangential to the rotor, $u\sin\theta_{yaw}$. If the blade at the top position in Figure 13.5, corresponding to $\theta_{wing} = 0°$, moves in the same direction as the wind, the relative rotational speed $\omega r(1+a')$ is reduced by $u\sin\theta_{yaw}$; at the bottom position, $\theta_{wing} = 180°$, it will move against the wind speed and the relative rotational speed is increased by $u\sin\theta_{yaw}$. Further, the axial induced velocity, and thus the axial velocity u, is not constant in an annular element of the rotor disc (see equation (9.26)).

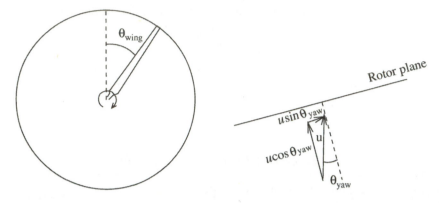

Figure 13.5 *Rotor plane showing the azimuthal position of a blade and a yawed rotor plane seen from the top*

For pure yaw the relative rotational speed varies sinusoidally as $\omega r(1+a') - u\sin\theta_{yaw}\cos\theta_{wing}$. Therefore both the relative velocity seen by the blade and the angle of attack vary with the frequency of the rotor, ω. The loads therefore also vary in yaw; this will therefore contribute to the fatigue loads and thus influence the expected lifetime of the rotor. It should now be clear that the rotor blades experience a variation of the angle of attack due to turbulence, wind shear, tower passage and yaw/tilt. The forces and moments on the blades and thus on the entire structure will therefore also vary in time. Since a wind turbine is designed to last at least 20 years, it is very important to quantify the loads in order to be able to perform a reliable fatigue analysis.

To make a wind turbine last for the design period, one must also take situations like extreme wind speeds into account. In this case the blades are parked or idling and the Danish standard (DS 412, 1992), for example, describes how to compute the extreme loads as:

$$p(r) = q_{2s}C_f c(r),\qquad\qquad(13.4)$$

where C_f is a force coefficient, $c(r)$ the chord and $q_{2s} = \frac{1}{2}\rho V_{2s}^2$ the dynamic pressure from an extreme wind speed time averaged over 2 seconds. For a Danish homogeneous terrain, V_{2s} can be computed using:

$$V_{2s} = V_b k_t (1n(\frac{h}{z_o}) + 3) = V_{10min} + 3\sigma,\qquad\qquad(13.5)$$

where $V_b = 27$m/s is a basis wind speed, h the height above the ground (minimum 4m), z_o the roughness length and k_t a terrain factor. σ is the standard deviation in a 10 minute time series. The terrain factor k_t is related to the roughness length z_o as:

z_o [m]	k_t
0.004	0.16
0.010	0.17
0.050	0.19
0.300	0.22

The extreme wind speed time, averaged over 10 minutes, can be estimated from:

$$V_{10min} = V_b k_t \ln(\frac{h}{z_o}).$$ (13.6)

The following simplified example shows how to quantify extreme loads on wind turbine blades. In the Danish standard (DS 472, 1992) for loads and safety in wind turbine construction, it is stated that for the blades the extreme wind speed V_{2s} must be calculated at $h = h_{hub}+2/3R$ using $C_f = 1.5$. For a wind turbine with a hub height of 40m and a rotor radius of 20m, this corresponds to $h = 40+2/3 \cdot 20 = 53.3$m. If the surrounding landscape has no nearby obstacles, such as houses, and very low vegetation, the roughness length is approximately 0.01m and the terrain factor $k_t = 0.17$. In this case $V_{2s} = 27 \cdot 0.17(\ln(53.3/0.01)+3) = 53.2$m/s. Assuming that the chord, c, is a constant 1.3m and the density is 1.28kg/m³, the load according to equation (13.4) is $p = \frac{1}{2} \cdot 1.28 \cdot 53.2^2 \cdot 1.3 \cdot 1.5 = 3532$N/m. The root bending moment at $r = 3$m for the constant load then becomes:

$$M = \int_r^R rp(r)dr = \frac{1}{2}p(R^2 - r^2) = \frac{1}{2}3532(20^2 - 3^2) = 690533 Nm.$$ (13.7)

For a simplified structural cross section such as the one shown in Figure 13.6, the moment of inertia about the flapwise axis is:

$$I = \frac{1}{12}a((2b_1)^3 - (2b_2)^3) = \frac{2}{3}a(b_1^3 - b_2^3).$$ (13.8)

Further it is assumed that $E = 14$Gpa $= 14 \times 10^9$N/m² and that the thickness of the aerofoil is $t/c = 35\%$. The thickness is thus $t = (t/c) \cdot c = 0.35 \cdot 1.3$m $= 0.455$m and $b_1 = t/2 = 0.228$m.

Given the extreme wind speed, the thickness of the aerofoil and E, it is possible to estimate the necessary shell thickness $t_s = b_1 - b_2$ so that the blade does not break.

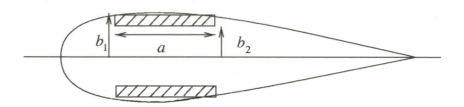

Figure 13.6 *Simplified structural model*

Since the force in the tangential direction is small, equation (11.10) reduces to:

$$\varepsilon = \frac{M}{EI}y \tag{13.9}$$

and it is clear that the maximum strain occurs for $y = b_1$. From testing it has been found that the material used in this example fails for $\varepsilon = \varepsilon_{fail} = 0.02$. The necessary moment of inertia can now be estimated from equation (13.9) as:

$$I = \frac{M}{E\varepsilon_{fail}}b_1 = \frac{690533}{14 \cdot 10^9 \cdot 0.02} 0.228 = 5.62 \cdot 10^{-4} \text{m}^4. \tag{13.10}$$

Using this necessary moment of inertia in equation (13.8), b_2 can be found for $a = 0.5$m:

$$b_2 = (b_1^3 - \frac{3I}{2a})^{1/3} = (0.228^3 - \frac{3 \cdot 5 \cdot 62 \cdot 10^{-4}}{2 \cdot 0.5})^{1/3} = 0.217\text{m}. \tag{13.11}$$

The necessary shell thickness is thus $t_s = b_1 N b_2 = 0.228-0.217 = 0.011$m = 1.1cm. In the standard (DS 472, 1992) it is further stated, however, that the above sketched method can only be applied if the construction is assumed not to auto-vibrate.

References

DS 472 (1992) Dansk Ingeniørforenings og ingeniørsammenslutningens Norm for 'Last og sikkerhed for vindmøllekonstruktioner' (in Danish), Danish standard

Troen, I. and Petersen, E. L. (1989) *European Wind Atlas*, Risø National Laboratory

14

Wind Simulation

To calculate realistic time series for the loads on a wind turbine construction as shown in Figure 10.2, for example, it is important to have as input a wind field with correct spatial and temporal variation, as sketched in Figure 13.4. This variation should include turbulence, wind shear and the effect on the wind from the tower. This chapter describes some of the statistical properties of atmospheric turbulence and how to model a 3-D wind field.

Wind Simulation at One Point in Space

A simple anemometer measures the wind speed at one point with a sample frequency of $f_s = 1/\Delta t$, where Δt is the time between two measurements. The output is a list of numbers, u_i $i = 1,..,N$, where the corresponding time is $t = 1\cdot\Delta t, 2\cdot\Delta t,...,N\cdot\Delta t$ The total time is $T = \Delta t \cdot N$ and the sample frequency is $f_s = N/T$. An example of such a time history is shown in Figure 14.1.

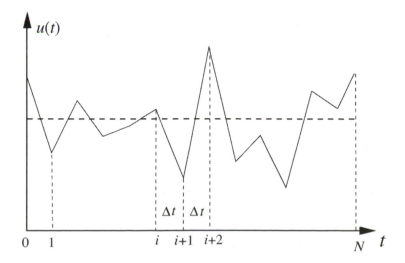

Figure 14.1 *Time history of discrete sampled wind speed at one point*

Figure 14.1 shows that the highest frequency that can be resolved is $f_h = f_s/2 = (N/2)/T$, since three discrete points are needed as a minimum to describe a vibration. The lowest frequency that can be resolved is $f_{low} = 1/T$. If the signal is assumed to be periodical, the time history can be decomposed using a discrete Fourier transformation as:

$$u(t) = a_o + \sum_{n=1}^{N/2} a_n \cos(\omega_n t) + b_n \sin(\omega_n t), \; \omega_n = \frac{2\pi n}{T} . \tag{14.1}$$

The coefficients are found as:

$$a_o = \frac{1}{N} \sum_{n=1}^{N} u_i \qquad \text{(mean wind speed)} \tag{14.2}$$

$$a_n = \frac{2}{N} \sum_{i=1}^{N} u_i \cos(\frac{2\pi n}{N} i), \; n = 1, \ldots, \frac{N}{2} - 1 \tag{14.3}$$

$$b_n = \frac{2}{N} \sum_{i=1}^{N} u_i \sin(\frac{2\pi n}{N} i), \; n = 1, \ldots, \frac{N}{2} - 1 \tag{14.4}$$

$$a_{\frac{N}{2}} = \frac{1}{N} \sum_{i=1}^{N} u_i \cos(\pi i) \tag{14.5}$$

$$b_{\frac{N}{2}} = 0 \tag{14.6}$$

Putting equation (14.1) into the definition of the variance, σ, the following relationship can be proved:

$$\sigma^2 = \frac{1}{N} \sum_{i=1}^{N} (u_i - \bar{u})^2 = \frac{1}{2} \sum_{n=1}^{N/2-1} (a_n^2 + b_n^2) + a_{\frac{N}{2}}^2. \tag{14.7}$$

For the power spectral density function, $PSD(f)$, the variance is:

$$\sigma^2 = \int_0^\infty PSD(f) df, \tag{14.8}$$

where f is the frequency. Discretizing the integral (14.8), and given that the frequencies lie between $f_{low} = 1/T$ and $f_{N/2} = (N/2)/T$, yields (see Figure 14.2):

$$\sigma^2 \approx \sum_{n=1}^{N/2} PSD(f_n) \Delta f, \; f_n = \frac{\omega_n}{2\pi} = \frac{n}{T}. \tag{14.9}$$

Comparing equations (14.9) and (14.7), it is seen that the power spectral density function $PSD(f)$ can be evaluated as:

$$PSD(f_n) = \frac{1}{2\Delta f}(a^2_n + b^2_n) = \frac{T}{2}(a^2_n + b^2_n). \tag{14.10}$$

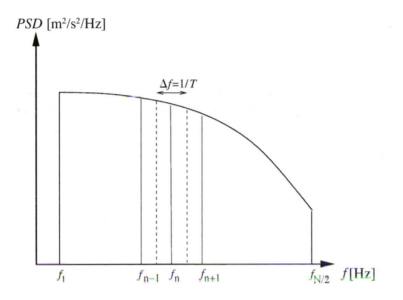

Figure 14.2 *Power spectral density function*

From a measured time series, $u_1,...,u_N$, the Fourier coefficients and thus the power spectral density function can be determined from equations (14.3), (14.4) and (14.10). This method is called discrete Fourier transform (DFT).

Constructing a time series from a known power spectral density (*PSD*) function is known as the inverse DFT. For the atmospheric boundary layer, different analytical expressions to approximate measured *PSD* functions exist, for example a Kaimal spectrum as given in DS 472 (1992):

$$PSD(f) = \frac{I^2 V_{10min} l}{(1+1.5\frac{f \cdot l}{V_{10mins}})^{5/3}} \tag{14.11}$$

$I = \sigma/V_{10min}$ is the turbulence intensity, f is the frequency (in Hz), V_{10min} is the 10 minute averaged wind speed, and l is a length scale, $l = 20h$ for $h < 30m$ and

$l = 600$m for $h > 30$m, where h is the height above ground level. The inverse DFT should provide a Fourier transform, as shown below, that satisfies the prescribed *PSD* function:

$$u(t) = \bar{u} + \sum_{n=1}^{N/2} a_n \cos(\omega_n t) + b_n \sin(\omega_n t). \tag{14.12}$$

Equation (14.12) can be rewritten as:

$$u(t) = \bar{u} + \sum_{n=1}^{N/2} \sqrt{a_n^2 + b_n^2} \cos(\omega_n t - \varphi_n), \tag{14.13}$$

where φ_n is the phase angle at frequency ω_n. Replacing the term $\sqrt{a_n^2 + b_n^2}$ by $\sqrt{2PSD(f_n)/T}$ in equation (14.13) using, for example, equation (14.11) for the spectrum, yields a Fourier transform that exactly satisfies the prescribed *PSD* function:

$$u(t) = \bar{u} + \sum_{n=1}^{N/2} \sqrt{\frac{2PSD(\omega_n)}{T}} \cos(\omega_n t - \varphi_n). \tag{14.14}$$

$$t = i \cdot \Delta t \text{ for } i = 1,...,N$$

The phase angle φ_n is not reflected in the *PSD* function and can be modelled using a random number generator yielding a value between 0 and 2π.

Using equation (14.14) it is very easy to compute a discrete time series having exactly the prescribed *PSD* function. A note should be made on the *PSD* function such as, for example, equation (14.11). These functions assume that all frequencies between 0 and ∞ are present, but as described earlier only frequencies between $1/T$ and $(N/2)/T$ are present for the discrete time series. It is therefore practical to scale the *PSD* function as:

$$\int_{f=1/T}^{f=(N/2)/T} PSD(f)df = 1, \tag{14.15}$$

corresponding to a standard deviation of 1, in other words a turbulent intensity of $I = 1/V_{10min}$. A standard deviation of S and thus a turbulent intensity of $I = S/V_{10min}$ can be found simply as:

$$u_i - \bar{u} = S(u_i - \bar{u}) \tag{14.16}$$

or:

$$u_i = Su_i + (1-S)\bar{u} \tag{14.17}$$

More information on different spectras such as the Kaimal and von Karman can be found in Burton et al (2001) and Rohatgi and Nelson (1994).

3-D Wind Simulation

To simulate a time history of the wind speed at two or more points in space, it must also be considered that the time histories are not independent. This dependency is of course related to the physical distance between two points, but also to the frequency. The high frequency content of the time series is a result of small vortices, which have small spatial influence. Similarly the low frequency part is related to large-scale vortices covering a bigger volume of the flow. A necessary coherence function must therefore take into account both the distance, L, between points j and k and the frequency; one possible choice is given in DS 472 (1992):

$$coh_{jk}(L,f) = \exp(-12(fL/V_{10min})).\qquad(14.18)$$

Veers (1988) presents a method that will generate a 3-D wind field with a prescribed *PSD* function and coherence function. In the following the method will not be proved but only given as an algorithm.

First, a matrix, S_{jk}, is created:

$$S_{jk} = coh_{jk}\sqrt{S_{jj}\cdot S_{kk}}.\qquad(14.19)$$

where S_{jj} and S_{kk} are the *PSD* functions of points j and k respectively. The off-diagonal terms are the magnitudes of the cross-spectras. If the number of points in space is *NP*, S_{jk} is an *NP*×*NP* matrix.

Second, a lower triangular **H** matrix is computed through following recursive formulae:

$$
\begin{aligned}
H_{11} &= S_{11}^{1/2}\\
H_{21} &= S_{21}/H_{11}\\
H_{22} &= (S_{22} - H_{21}^2)^{1/2}\\
H_{31} &= S_{31}/H_{11}\\
&\vdots\\
H_{jk} &= (S_{jk} - \sum_{l=1}^{k-1} H_{jl}H_{kl})/H_{kk}\\[1em]
H_{kk} &= (S_{kk} - \sum_{l=1}^{k-1} H_{kl}^2)^{1/2}
\end{aligned}
\qquad(14.20)
$$

For each point indexed by k and for each discrete frequency, $f_m = m/T$, a random number, φ_{km}, between 0 and 2π is found to represent the phase as in equation (14.14). m varies between 1 and $N/2$, where N is the number of discrete points in the time histories ($t = i \cdot \Delta t, i = 1,\ldots,N$).

A vector with imaginary components of length equal to the number of points in space, $\mathbf{V} = V_j(f_m)$, is now calculated as:

$$\text{Re}(V_j(f_m)) = \sum_{k=1}^{j} H_{jk}\cos(\varphi_{km})$$

$$\text{Im}(V_j(f_m)) = \sum_{k=1}^{j} H_{jk}\sin(\varphi_{km})$$

(14.21)

$\text{Re}(V_j(f_m))$ and $\text{Im}(V_j(f_m))$ is transformed to an amplitude, $Amp_j(f_m)$, and a phase, $\Phi_j(f_m)$, as:

$$Amp_j(f_m) = \sqrt{\text{Re}(V_j(f_m))^2 + \text{Im}(V_j(f_m))^2}$$

(14.22)

$$\tan\Phi_j(f_m) = \frac{\text{Im}(V_j(f_m))}{\text{Re}(V_j(f_m))}$$

Finally, the time histories at the points $j = 1,NP$ can be computed as:

$$U_j(t) = \bar{U} + \sum_{m=1}^{N/2} 2Amp_j(f_m)\cos(2\pi f_m \cdot t - \Phi_j(f_m))$$

(14.23)

$$t = i \cdot \Delta t \text{ for } i = 1,\ldots,N$$

Figure 14.3 plots the time series of two points spaced 1m apart. The mean wind speed is 10m/s and the turbulence intensity is 0.1. It is clearly seen that the two curves are well correlated for the lower frequencies. Figure 14.4 plots the coherence function computed from the two time series shown in Figure 14.3 together with the input given by equation (14.18).

Using equation (14.23), together with appropriate *PSD* and coherence functions, the time histories at all points are computed for each velocity component $U = (u,v,w)$ independently. Therefore there is no guarantee of obtaining correct cross-correlations. In Mann (1998) a method ensuring this is developed on the basis of the linearized Navier-Stokes equations. In the future, wind fields are expected to be generated numerically from large eddy simulations (LES) or direct numerical simulations (DNS) of the Navier-Stokes equations for the flow on a landscape similar to the actual siting of a specific wind turbine.

For an aeroelastic calculation of a wind turbine construction it is natural to construct the time histories in a series of points arranged as indicated in

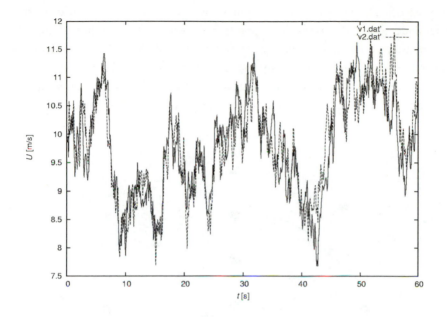

The mean wind speed is 10m/s and the turbulence intensity is 0.1.

Figure 14.3 *Computed time series of wind speed in two points separated by 1m*

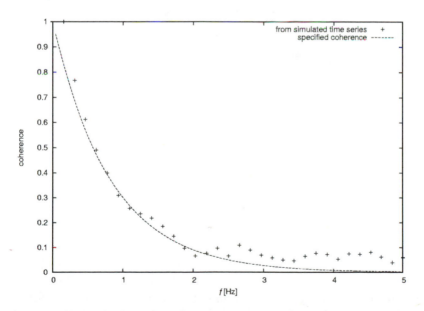

Figure 14.4 *Comparison of actual coherence from the two time series shown in Figure 14.3 and specified by equation (14.18)*

Figure 14.5. The velocities on the blade sweeping through the grid must in general be found by spatial interpolation. It should be mentioned that the time history of the wind seen by a point on the blade is different from the time history of a point fixed in space. A time history for a point on the rotating blade is called rotational sampling; Veers (1988) shows how this can be directly calculated for a blade with a constant rotational speed.

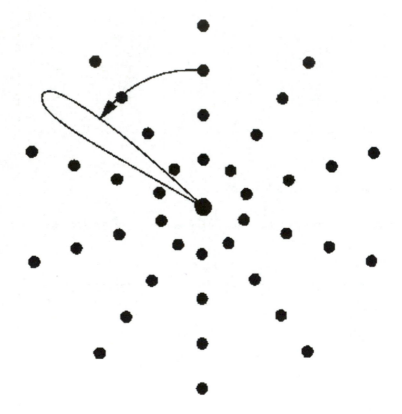

Figure 14.5 *Sketch of point distribution for time histories of wind speed for aeroelastic calculation of wind turbine construction*

To illustrate rotational sampling a number of time series are created for a constant radius of $r = 5m$ and azimuthally spaced by 7.5°. The velocity seen by a blade at the same radius and rotating with $\omega = 13.1$ rad/s is found by interpolation in the time series. In Figure 14.6 the calculated *PSD* from the rotational sampling is plotted together with the *PSD* from a stationary point

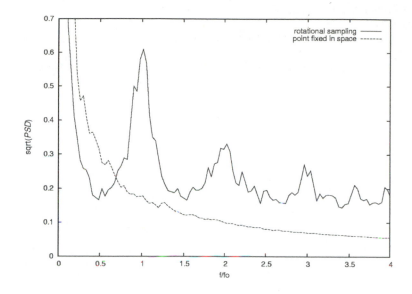

Figure 14.6 *Comparison between specified* PSD *and the* PSD *seen by rotating blade*

in space. The *PSD* for the rotating blade shows clear spikes at multiples of the rotational frequency 1P, 2P,... (the frequency in Figure 14.6 is non-dimensionalized with the rotational frequency of the rotor, $f_0 = \omega/2\pi$). These spikes, which will make a major contribution to the fatigue damage of the wind turbine construction, are due to the spatial coherence, since for a frozen time there exist areas of the rotor disc where the velocity is relatively high or low (see, for example Figure 13.4), and a blade will pass these areas once per revolution.

Finally, it should be noted that the simulated time series not only depend on the prescribed *PSD* and coherence functions, but also on the random numbers for the phases. Therefore it is recommended to calculate at least three different time series for each case and perform for each series a run with the aeroelastic code to get an idea of the uncertainty from the different time series.

References

Burton, T., Sharpe, D., Jenkins, N. and Bossanyi, E. (2001) *Wind Energy Handbook*, John Wiley & Sons, Chichester, UK

DS 472 (1992) Dansk Ingeniørforenings og ingeniørsammenslutningens Norm for 'Last og sikkerhed for vindmøllekonstruktioner' (in Danish), Danish standard

Mann, J. (1998) 'Wind field simulation', *Problems in Engineering Mechanics*, vol 13, pp269–282

Rohatgi, J. S. and Nelson, V. (1994) *Wind Characteristics*, West Texas A&M University, Canyon, TX

Veers, P. (1988) 'Three-dimensional wind simulation', SANDIA report, SAND88-0152 UC-261,Sandia National Laboratories, USA

15

Fatigue

From earlier chapters it is clear that the loads on a wind turbine vary constantly with time, giving rise to a possible breakdown due to accumulated fatigue damage. In Madsen et al (1990) a recommended practice to estimate the fatigue damage, and thus the lifetime of a wind turbine, is outlined. This chapter provides a summary of this practice.

First, the loads must be obtained from either computations using an aeroelastic code or directly from measurements. For normal operation the loads are monitored for 10 minutes in each wind speed interval $V_p < V_o < V_{p+1}$. An example of such a time history for $V_{10min} = 11$m/s and a turbulence intensity $I = 0.1$ is seen in Figure 10.2, which is a result of a simulation using the aeroelastic code FLEX. The turbulence intensity, I, is defined as σ/V_{10min}, where σ is the standard deviation of the wind speed within the 10 minute time series. Knowing the loads, the stresses at critical points on the wind turbine are computed using equations (11.9) (Hook's law) and (11.10). As a minimum it is recommended in Madsen et al (1990) to monitor the blade bending moments, the yaw and tilt rotor moments, the axial thrust, the torque in the main shaft, the bending moments of the tower, and the torsional moment in the tower. From each 10 minutes time history the stresses are sorted in a matrix, where the elements $m_{ij}(V_p < V_o < V_{p+1})$ denote the number of cycles in the mean stress interval $\sigma_{m,i} < \sigma_m < \sigma_{m,i+1}$ and range interval $\sigma_{r,j} < \sigma_r < \sigma_{r,j+1}$ for the wind speed interval $V_p < V_o < V_{p+1}$. Figure 15.1 sketches one cycle with a mean stress value σ_m and a range σ_r; it is seen that the range is twice the amplitude σ_a.

To count the number of cycles from an actual time series such as the one shown in Figure 10.2, a technique called 'rainflow counting' is used (for a complete description of this algorithm, see Madsen et al, 1990). Then, knowing the annual wind distribution $h_w(V_o)$, the probability, f, of the wind speed being in the interval $V_p < V_o < V_{p+1}$ is computed from equation (6.48). The actual number of annual 10 minute periods where the wind speed is in this interval is $6 \cdot 8760 \cdot f$. The number of cycles per year, n_{ij}, in the mean stress interval $\sigma_{m,i} < \sigma_m < \sigma_{m,i+1}$ and in the range interval $\sigma_{r,j} < \sigma_r < \sigma_{r,j+1}$ is found by

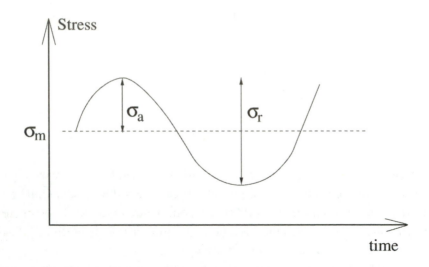

Figure 15.1 *Definition of mean stress σ_m and range σ_r for one cycle*

adding together the contributions from each wind speed interval:

$$n_{ij} = \sum_{p=1}^{N-1} m_{ij}(V_p < V_o < V_{p+1}) \cdot 6 \cdot 8760 \cdot f(V_p < V_o < V_{p+1}), \tag{15.1}$$

where $N-1$ is the number of wind speed intervals. The matrix **M**, with elements n_{ij}, is called the Markov matrix. A wind turbine also experiences loads when starting and stopping and when running under abnormal conditions such as at high yaw angles. Before a lifetime analysis can be performed, it is also necessary to estimate these loads with respect to the annual number of occurrences with a given mean stress interval $\sigma_{m,i} < \sigma_m < \sigma_{m,i+1}$ and range interval $\sigma_{r,j} < \sigma_r < \sigma_{r,j+1}$ and add this to the Markov matrix n_{ij}. In so doing one has to distinguish between starting and stopping at high wind speed and low wind speed, since the loads are different in these two cases. IEC 61400 (2004) gives a complete list of the different load cases needed for certification. The total number of cycles in the entire lifetime within the mean stress interval $\sigma_{m,i} < \sigma_m < \sigma_{m,i+1}$ and range interval $\sigma_{r,j} < \sigma_r < \sigma_{r,j+1}$ is:

$$n_{tot,if} = T \cdot n_{ij} \tag{15.2}$$

where the lifetime T is measured in years. To estimate T, the Palmgren-Miner rule, equation (15.3), for cumulative damage during cyclic loading is used. This rule assumes that the ratio between the number of applied stress cycles,

n_{ij}, with a given mean stress level $\sigma_{m,i}$ and range $\sigma_{r,j}$, and the number of cycles, N_{ij}, which with the same mean stress and range would lead to failure, constitutes the expended part of the useful fatigue life and that the sum of these ratios is thus the damage D. Thus the criteria for not failing is that D is less than 1:

$$\sum \frac{n_{tot,ij}}{N_{ij}} = D < 1 \tag{15.3}$$

Combining equation (15.2) with equation (15.3) yields the following equation to estimate the lifetime T:

$$T = \frac{1}{\sum \dfrac{n_{ij}}{N_{ij}}} \tag{15.4}$$

The number of cycles, N_{ij}, leading to failure for a cyclic loading with a given mean stress level $\sigma_{m,i}$ and range $\sigma_{r,j}$ for a given material is found in a so-called *S-N* curve or Wöhler curve like the one sketched in Figure 15.2.

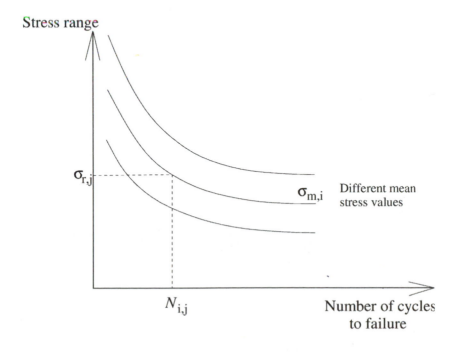

Figure 15.2 *Sketch of an* S-N *curve*

Sometimes only the range $\sigma_{r,j}$ is taken into account and the influence of the mean stress level $\sigma_{m,i}$ is ignored. Under this assumption the *S-N* curve can be approximated by:

$$N = N^* \left(\frac{\sigma_r(N^*)}{\sigma_r} \right)^m \tag{15.5}$$

where m is a material constant and $\sigma_r(N^*)$ is the stress range giving failure for N^* cycles. According to DS 412 (1983), for steel m is approximately 4.0; for glass fibre m is approximately 8–12.

The damage D can be estimated using the Palmgren-Miner rule. To compare the contribution from the different wind speeds to the total fatigue damage, an equivalent, $\sigma_{r,eq}$, load can be used. The equivalent load is defined as the cyclic load which, when applied n_{eq} times, gives the same fatigue damage on the wind turbine as the real turbulent flow at the considered wind speed. Since the total damage, D, is known, the equivalent load can be calculated similarly to using equation (15.5) for the *S-N* curve:

$$D = \frac{n_{eq}}{N_{eq}} = \frac{n_{eq}}{N^* \left(\frac{\sigma_r(N^*)}{\sigma_{r,eq}} \right)^m} \tag{15.6}$$

$$\Downarrow$$

$$\sigma_{r,eq} = \sigma_r(N^*) \left(\frac{N^* D}{n_{eq}} \right)^{1/m}$$

Figure 15.3 plots a time series of 300s of a flapwise bending moment; Figure 15.4 shows the result of using rainflow counting on this signal; in other words, the number of cycles with a given range contained in the time series.

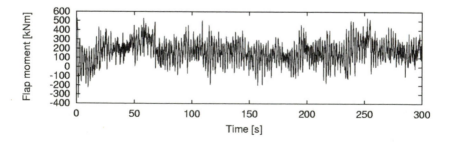

Figure 15.3 *An example of a time history of a flapwise bending moment*

Number of cycles

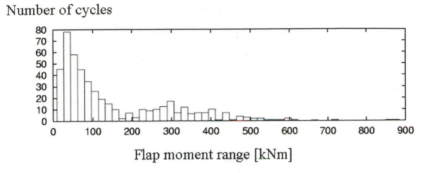

Flap moment range [kNm]

Figure 15.4 *Result of using rainflow counting on the time series from Figure 15.3*

A time series consisting of a sequence of cyclic loads with increasing range can be made that, assuming the Palmgren-Miner rule is correct, will give the same fatigue damage as the original time series. Such a time series is shown in Figure 15.5.

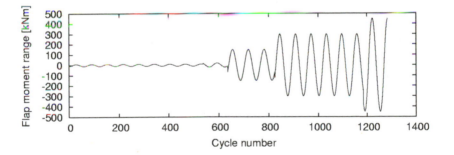

Figure 15.5 *Sequence of cyclic loads with increasing range that gives the same fatigue damage as the original time series*

References

DS 412 (1983) Dansk Ingeniørforenings norm for stålkonstruktioner (Danish standard for steel constructions)

IEC 61400-1 (2004) 'Wind turbines. Part 1: Design requirements', CD, edition 3, second revision, IEC TC88-MT1

Madsen, P. H., Dekker, J. W. M., Thor, S. E., McAnulty, K., Matthies, H. and Thresher, R. W. (1990) 'Expert group study on recommended practices for wind turbine testing and evaluation. 3: Fatigue loads', IEA Wind Energy Conversion Systems, second edition

16

Final Remarks

This book has hopefully provided an insight into wind turbine aerodynamics and aeroelasticity. To model the steady performance of a given wind turbine in order to calculate the annual energy production for an actual site, the classical steady BEM method has been outlined. The loads on a real wind turbine construction are, however, very dynamic, mainly due to gravity and the varying inflow conditions from wind shear, tower shadow and, not least, atmospheric turbulence. All the theory and equations needed to write an unsteady BEM code are given in this book, including all necessary engin-eering models, such as dynamic inflow, dynamic stall and yaw/tilt models. A method to build a structural model of a wind turbine construction has also been outlined in order to allow the calculation of the dynamic structural response of the different components when exposed to unsteady loads. The vibrations couple directly to the aerodynamic loads through the angle of attack since, when estimating the local velocities seen by the blade, the velocities from the vibrations must be subtracted from the wind speed. The aerodynamics and structural dynamics are therefore strongly coupled and comprise a so-called aeroelastic problem in which both models must be solved simultaneously in a time marching procedure. During an aeroelastic simulation the aerodynamic model must be solved many times (one time per time step) and a fast model is thus required. The BEM method is simple but very fast and will therefore very likely be used for many years to come. However, more advanced tools such as CFD (computational fluid dynamics) have made huge progress in recent years, not least due to advances in processor speed and storage capacity of modern computers.

In CFD the domain around the object is divided into a computational grid and for each grid point the Navier-Stokes equations (A10–A13) are discretized; the result is a large number of coupled non-linear equations that must be solved numerically, and, further, if an unsteady solution is sought, they must be solved for each time step. Another problem is the variety of scales present in the actual flow, ranging from tiny turbulent eddies to large scales in the order of the rotor diameter. Solving the equations in a computational grid that resolves all the scales will remain impossible for many years to come since the size of the equation system required is enormous. To overcome this, the smallest eddies are modelled using a turbulence model that basically models the extra mixing from the small-scale turbulence. The existing turbulence models are often calibrated for various different flows and therefore their use introduces additional uncertainty to the computed results. One of the very first papers using CFD on wind turbines was Hansen et al (1997). More recent calculations (Sørensen and Michelsen, 2000; Duque et al, 1999) of the National Renewable Energy Laboratory (NREL) experiment described in Fingersh et al (2001) and Simms et al (2001) have shown excellent agreement between the experiment and the CFD computations. Even though CFD has improved, however, it is still far too slow to be used in an aeroelastic computation and wind turbines will for many years still be designed and optimized using BEM codes. Nevertheless, it is very natural to check the final result using CFD to validate the design and to see if, for example, there are unwanted areas of separation. Figure 16.1 (from Hansen and Johansen, 2004) shows an example of the computed flow past a wind turbine blade equipped with two different tips.

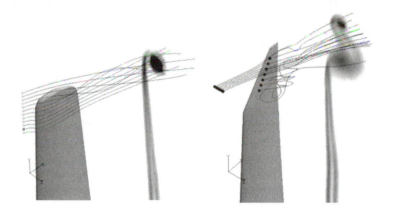

Figure 16.1 *Computed streamlines past two different tips on the same blade and iso-vorticity in a plane just behind the blade for the same rotational speed and wind speed*

In the so-called 'actuator line' model (Sørensen and Shen, 2002), CFD is used to resolve the wake dynamics, and thus the induced velocity at the rotor plane can also be evaluated. Knowing the induced velocities, the local angles of attack can be estimated just as in the BEM approach; the loads can then be looked up in a table of tabulated aerofoil data. Using the actuator line model, the effect from yaw and dynamic inflow is a direct output from the Navier-

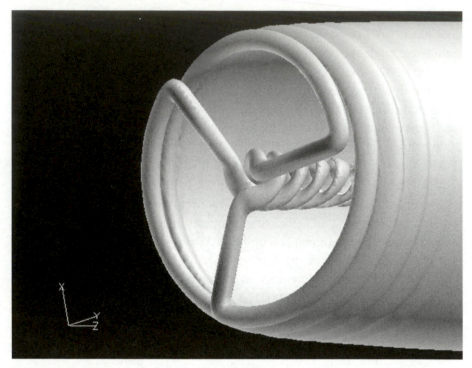

Figure 16.2 *Iso-vorticity plot of the flow past a wind turbine modelled with the actuator line model*

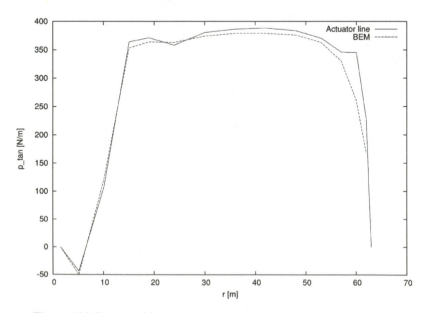

Figure 16.3 *Tangential loads computed using BEM and the actuator line model*

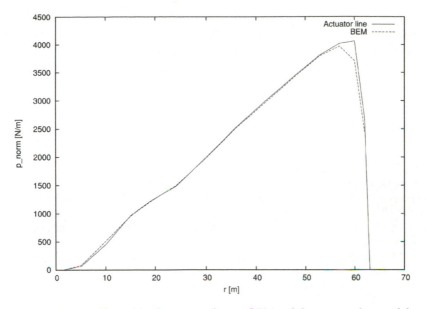

Figure 16.4 *Normal loads computed using BEM and the actuator line model*

Stokes equations and the results can be used to tune the engineering models in the BEM method. In Figure 16.2 the flow past a wind turbine simulated using the actuator line model is shown as an iso-vorticity plot. Figures 16.3 and 16.4 show the resulting tangential and normal loading for the same case as shown in Figure 16.2 (zero yaw and tip speed ratio of $\lambda = 7.6$) using a BEM code and the actuator line model, and a very good agreement is seen.

Further, the actuator line model can be used to calculate the wake effect from upstream wind turbines, since the wakes generated from these hit the downwind turbines (see Figure 16.5).

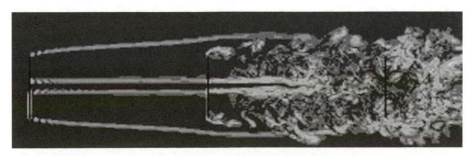

Figure 16.5 *Iso-vorticity plot of the flow past three wind turbines placed in a row aligned with the incoming wind*

The author wishes good luck to all those who decide to pursue the task of using and developing aeroelastic models for wind turbines in order to harvest more efficiently the vast energy resources contained in the wind.

References

Duque, E. P. N., van Dam, C. P. and Hughes, S. (1999) 'Navier-Stokes simulations of the NREL combined experiment phase II rotor', in *Proceedings of the ASME Wind Energy Symposium*, Reno, NV

Fingersh, L. J., Simms, D., Hand, M., Jager, D., Contrell, J., Robinson, M., Schreck, S. and Larwood, S. (2001) 'Wind tunnel testing of NREL's unsteady aerodynamics experiment', AIAA-2001-0035 Paper, 39th Aerospace Sciences Meeting and Exhibit, Reno, NV

Hansen, M. O. L. and Johansen, J. (2004) 'Tip studies using CFD and comparison with tip loss models', *Wind Energy*, vol 7, pp343–356

Hansen, M. O. L., Sørensen, J. N., Michelsen, J. A. and Sørensen, N. N. (1997) 'A global Navier-Stokes rotor prediction model', AIAA 97-0970, 35th Aerospace Sciences Meeting & Exhibit, Reno, NV

Simms, D., Schreck, S., Hand, M. and Fingersh, L. J. (2001) 'NREL unsteady aerodynamics experiment in the NASA-Ames wind tunnel: A comparison of predictions to measurements', NREL/TP-500-29494, National Renewable Energy Laboratory

Sørensen, J. N. and Shen, W. Z. (2002) 'Numerical modeling of wind turbine wakes', *Journal of Fluids Engineering – Transactions of the ASME*, vol 124, no 2, pp393–399

Sørensen, N. N. and Michelsen, J. A. (2000) 'Aerodynamic predictions for the unsteady aerodynamics experiment phase-II rotor at the National Renewable Energy Laboratory', AIAA-2000-0037, 38th Aerospace Sciences Meeting & Exhibit, Reno, NV

Appendix A

Basic Equations in Fluid Mechanics

In a fluid with no individual solid particles it is common to consider a fixed volume in space, denoted as a control volume (CV). Newton's second law is:

$$\mathbf{F} = \frac{d\mathbf{P}}{dt}, \tag{A.1}$$

where $\mathbf{F} = (F_x, F_y, F_z)$ is the total force, $\mathbf{P}$ is the momentum and t is the time. The time derivative of the momentum $\mathbf{P}$ is found from integrating over the control volume as:

$$\frac{d\mathbf{P}}{dt} = \frac{\partial}{\partial t} \iiint_{CV} \rho \mathbf{V} d(Vol) + \iint_{CS} \mathbf{V} \rho \mathbf{V} \cdot d\mathbf{A}, \tag{A.2}$$

where ρ is the density, $\mathbf{V}$ is the velocity, $d(Vol)$ is an infinitesimal part of the total control volume, CS denotes the surface of the control volume and $d\mathbf{A}$ is a normal vector to an infinitesimal part of the control surface. The length of $d\mathbf{A}$ is the area of this infinitesimal part. Newton's second law for the control volume then becomes:

$$\frac{\partial}{\partial t} \iiint_{CV} \rho \mathbf{V} d(Vol) + \iint_{CS} \mathbf{V} \rho \mathbf{V} \cdot d\mathbf{A} = \mathbf{F}, \tag{A.3}$$

where $\mathbf{F}$ is the total external force including the pressure and viscous forces acting on the control surfaces. Further, body forces, for example gravity, and forces from the flow past an object inside the control volume contribute to the total force. Equation (A.3) is normally used to determine an unknown force, provided that the velocity is known at the control surfaces. When Stoke's hypothesis for an incompressible fluid, equations (A.4–A.9), is used for the stresses on an infinitesimal control volume with side lengths (dx, dy, dz), the three partial differential momentum equations (A.11), (A.12) and (A.13) are derived. The first subscript on τ indicates the face where the stress is located; the second subscript is the direction of the stress:

$$\tau_{xx} = -p + 2\mu \frac{\partial u}{\partial x} \tag{A.4}$$

$$\tau_{xy} = \tau_{yx} = \mu\left(\frac{\partial u}{\partial y} + \frac{\partial v}{\partial x}\right) \tag{A.5}$$

$$\tau_{xz} = \tau_{zx} = \mu\left(\frac{\partial u}{\partial z} + \frac{\partial w}{\partial x}\right) \tag{A.6}$$

$$\tau_{yy} = -p + 2\mu \frac{\partial v}{\partial y} \tag{A.7}$$

$$\tau_{yz} = \tau_{zy} = \mu\left(\frac{\partial v}{\partial z} + \frac{\partial w}{\partial y}\right) \tag{A.8}$$

$$\tau_{zz} = -p + 2\mu \frac{\partial w}{\partial z} \tag{A.9}$$

$p(x,y,z,t)$ denotes the pressure, $\mathbf{V}(x,y,z,t) = (u,v,w)$ are the velocity components, $\mathbf{x} = (x,y,z)$ are the coordinates in a cartesian frame of reference and μ is the viscosity.

The three momentum equations (A.11), (A.12) and (A.13) plus the continuity equation (A.10) comprise the the Navier-Stokes equations for an incompressible fluid with constant viscosity μ:

$$\frac{\partial u}{\partial x} + \frac{\partial v}{\partial y} + \frac{\partial w}{\partial z} = 0 \tag{A.10}$$

$$\rho\left(\frac{\partial u}{\partial t} + u\frac{\partial u}{\partial x} + v\frac{\partial u}{\partial y} + w\frac{\partial u}{\partial z}\right) = -\frac{\partial p}{\partial x} + \mu\left(\frac{\partial^2 u}{\partial x^2} + \frac{\partial^2 u}{\partial y^2} + \frac{\partial^2 u}{\partial z^2}\right) + f_x \tag{A.11}$$

$$\rho\left(\frac{\partial v}{\partial t} + u\frac{\partial v}{\partial x} + v\frac{\partial v}{\partial y} + w\frac{\partial v}{\partial z}\right) = -\frac{\partial p}{\partial y} + \mu\left(\frac{\partial^2 v}{\partial x^2} + \frac{\partial^2 v}{\partial y^2} + \frac{\partial^2 v}{\partial z^2}\right) + f_y \tag{A.12}$$

$$\rho\left(\frac{\partial w}{\partial t} + u\frac{\partial w}{\partial x} + v\frac{\partial w}{\partial y} + w\frac{\partial w}{\partial z}\right) = -\frac{\partial p}{\partial z} + \mu\left(\frac{\partial^2 w}{\partial x^2} + \frac{\partial^2 w}{\partial y^2} + \frac{\partial^2 w}{\partial z^2}\right) + f_z \tag{A.13}$$

Equation (A.10) ensures that the net mass flow is zero in and out of an infinitesimal box with side lengths dx, dy, dz. Equations (A.11–A.13) are Newton's second law, in the x, y and z direction respectively, for an infinitesimal box in the fluid, which is fixed in space. The left hand side terms are the inertial forces and the right hand side terms are the pressure forces, the viscous forces and the external body forces $\mathbf{f}(t,x,y,z) = (f_x, f_y, f_z)$ acting on the box respectively. Equations (A.11–A.13) can also be written in vector notation as:

$$\rho\left(\frac{\partial \mathbf{V}}{\partial t} + (\mathbf{V} \cdot \nabla)\mathbf{V}\right) = -\nabla p + \mu\nabla^2\mathbf{V} + \mathbf{f}. \qquad (A.14)$$

If no external forces are present and if the flow is stationary and the viscous forces are zero, equation (A.14) reduces to:

$$-\frac{\nabla p}{\rho} = (\mathbf{V} \cdot \nabla)\mathbf{V} = \tfrac{1}{2}\nabla(\mathbf{V} \cdot \mathbf{V}) - \mathbf{V} \times (\nabla \times \mathbf{V}). \qquad (A.15)$$

The last equality in equation (A.15) comes from a vector identity. If the flow is irrotational, i.e. $\nabla \times \mathbf{V} = 0$, the Bernoulli equation (A.16) comes directly from equation (A.15) and is valid between any two points in the flow domain:

$$p + \tfrac{1}{2}\rho(u^2 + v^2 + w^2) = const. \qquad (A.16)$$

If the flow is not irrotational, it can be shown from equation (A.15) that the Bernoulli equation (A.16) is still valid, but only along a streamline.

To use the Bernoulli equation it is necessary that the flow is stationary, that no external forces are present and that the flow is incompressible and frictionless. The Bernoulli equation is generally valid along a streamline, but if the flow is irrotational, the equation is valid between any two points.

The Navier-Stokes equations are difficult to solve and often the integral formulation equation (A.3) is used in engineering problems. If the flow is stationary and the torque on the sides of an annular control volume is zero, the integral moment of momentum becomes:

$$\mathbf{M} = \iint_{CS} \mathbf{r} \times \mathbf{V}\rho\mathbf{V} \cdot \mathbf{dA}, \qquad (A.17)$$

where $\mathbf{M}$ is an unknown torque acting on the fluid in the control volume and $\mathbf{r}$ is the radius from the cylindrical axis. If the flow is uniform at the inlet and exit of the control volume and the only non-zero component of $\mathbf{M}$ is in the flow direction z, Euler's turbine equation (A.18) can be derived from equation (A.17):

$$P = M_z\omega = \omega\dot{m}(r_1 V_{\theta,1} - r_2 V_{\theta,2}). \qquad (A.18)$$

P is power removed from the flow on a mechanical shaft, ω is the rotational speed of the shaft, V_θ is the tangential velocity component, $\dot{m}$ is the mass flow through the control volume, and subscripts 1 and 2 denote the inlet and exit of the control volume respectively.

Another important equation is the integral conservation of energy or the first law of thermodynamics for a control volume, which for steady flow is:

$$P + Q = \iint\limits_{CS} \left(u_i + \frac{p}{\rho} + \tfrac{1}{2}(u^2 + v^2 + w^2) \right) \rho \mathbf{V} \cdot d\mathbf{A}, \tag{A.19}$$

where *P* and *Q* are the mechanical power and the rate of heat transfer added to the control volume and u_i is the internal energy.

Appendix B:

Symbols

A	rotor area, scaling factor
$\mathbf{a_{AB}}$	transformation matrix from system A to B
a	axial induction factor, tower radius
a'	tangential induction factor
a_n	Fourier coefficient
B	number of blades
b_n	Fourier coefficient
$\mathbf{C}$	damping matrix
C_p	power coefficient
C_T	thrust coefficient
C_l	lift coefficient
C_d	drag coefficient
C_m	moment coefficient
C_n	normal load coefficient
C_t	tangential load coefficient
C_θ	azimuthal component of axial velocity
c	chord
D	rotor diameter, drag
$\mathbf{dA}$	normal vector to area
E	modulus of elasticity
ED	moment of centrifugal stiffness
EI	moment of stiffness inertia
ES	moment of stiffness
$\mathbf{F}$	force (vector)
$\mathbf{F_g}$	generalized force (vector)
F	force, Prandtl's tip loss correction
$\mathbf{f}$	external body force (vector)
f	force, probability, frequency
f_s	separation function
f_n	frequency in discrete Fourier transformation
GI	torsional stiffness

H	tower height, form factor
h	height above ground level
h_w	Weibull distribution
I	moment of inertia, turbulence intensity
$\mathbf{K}$	stiffness matrix
k	form factor
k_t	terrain factor
L	lift, distance between two points in space
l	length scale
$\mathbf{M}$	torque (vector), mass matrix
M	torque, aerodynamic moment
M_G	generator torque
M_{flap}	flapwise bending moment
Ma	Mach number
$\dot{m}$	mass flow
m	mass per length
n	rotational speed of shaft
$\mathbf{P}$	momentum (vector)
P	power
PSD	power spectral density function
p	pressure, load
p_N	load normal to rotor plane
p_T	load tangential to rotor plane
p_c	centrifugal load
Q	rate of heat transfer
Re	Reynolds number
R	rotor radius, resistance
$\mathbf{r}$	radius (vector)
r	radius
SL	slip
S_{ij}	coherence function
T	thrust, total time
t	time
U	boundary layer edge velocity
u	x-component of velocity vector, axial velocity at rotor plane, deflection
u_1	velocity in wake
u_i	internal energy
$\dot{u}$	structural velocity
$\ddot{u}$	structural acceleration

u^{1f}	flapwise eigenmode deflection first
u^{1e}	edgewise eigenmode deflection first
u^{2f}	flapwise eigenmode deflection second
$\mathbf{V}$	velocity (vector)
$\mathbf{V_b}$	blade velocity (vector)
V_o	wind speed
V_∞	velocity at infinity
V_{rel}	relative velocity to aerofoil
V_θ	tangential velocity component
V_2	velocity in rotor plane for a shrouded rotor
V_{10min}	time averaged wind speed over a period of 10 minutes
V_{2s}	time averaged wind speed over a period of 2 seconds
v	angle between chord line and first principal axis
ν	y-component of velocity vector
w	z-component of velocity vector
$\mathbf{w}$	induced velocity
W_y	tangential component of induced velocity
W_z	normal component of induced velocity
q_{2s}	dynamic pressure based on V_{2s}
x	local tip speed ratio
z_o	roughness length

Greek Characters

α	angle of attack
β	twist of blade
Γ	circulation
Δt	time increment
δ	boundary layer thickness
δ^*	displacement thickness
ε	augmentation factor, strain
θ	momentum thickness, local pitch
θ_{cone}	cone angle
θ_p	pitch angle
θ_o	azimuthal position where blade is deepest into the wake
θ_{wing}	azimuthal position of blade
θ_{yaw}	yaw angle
κ	curvature about the principal axis
λ	tip speed ratio
μ	dynamic viscosity

ν	kinematic viscosity, wind shear exponent
ρ	density
σ	solidity, stress, standard deviation
σ_r	stress range
σ_m	mean stress
τ	shear stress, time constant
ϕ	flow angle
χ	wake skew angle
ω	angular velocity of rotor, eigenfrequency
ω_n	frequency in discrete Fourier transformation

Index